BestMasters

Springer awards „BestMasters" to the best master's theses which have been completed at renowned universities in Germany, Austria, and Switzerland.

The studies received highest marks and were recommended for publication by supervisors. They address current issues from various fields of research in natural sciences, psychology, technology, and economics.

The series addresses practitioners as well as scientists and, in particular, offers guidance for early stage researchers.

Johannes Schötz

Attosecond Experiments on Plasmonic Nanostructures

Principles and Experiments

With a Preface by Prof. Dr. Matthias Kling

Johannes Schötz
Garching, Germany

BestMasters
ISBN 978-3-658-13712-0   ISBN 978-3-658-13713-7 (eBook)
DOI 10.1007/978-3-658-13713-7

Library of Congress Control Number: 2016937504

Springer Spektrum

Printed on acid-free paper

This Springer Spektrum imprint is published by Springer Nature
The registered company is Springer Fachmedien Wiesbaden GmbH

To my family, friends and colleagues

Preface

Attosecond nanophysics is a new research field merging ultrafast science with time scales reaching into the attosecond domain with studies on nanoscale materials. An attosecond is incredibly short. To put it in perspective, one attosecond (one attosecond $= 10^{-18}$ seconds) compares to one second roughly as one second compares to the age of the universe. Within one attosecond even light only travels a distance of 0.3 nanometer (1 nanometer $= 10^9$ meter). Attosecond time and nanometer length scales are thus inherently connected. The attosecond timescale is particularly important for electrons, which are light enough to move so fast that they must be clocked with attosecond precision to track their motion. These fast electron dynamics govern the interaction of light with matter and form the basis for optoelectronics. The possibility to steer electronic processes in nanomaterials with tailored lightwaves can be exploited in ultrafast nanoelectronic circuitry with switching frequencies approaching the petahertz domain (many orders of magnitudes above conventional electronics). This potential has motivated the rapid growth of attosecond nanophysics.

The master thesis of Johannes Schötz discusses an important experimental advance in this young field, namely the ability to measure the evolution of fields on the nanoscale in real-time, i.e. attosecond timescales. He describes both experimental and theoretical advances towards the realization of the attosecond streak-camera technique on the nanoscale. While the attosecond streak-camera has become a standard tool in attosecond physics, and related measurements on electron dynamics in atoms, molecules, and extended surfaces, its realization for measurements of nanostructures is not straightforward.

The reasons are discussed in detail in the thesis, with a special emphasis on metallic nanotips, which Johannes Schötz has investigated in his work.

In attosecond streaking, electrons are photoemitted through an attosecond light pulse in the extreme ultraviolet and are accelerated by an external field provided by e.g. a synchronized optical light pulse (with a duration of a few cycles). While in conventional attosecond streaking the external fields are spatially homogenous, the near-fields of nanostructures are inhomogenous. The ramifications of the nanometer spatial inhomogeneity are non-trivial and therefore typically detailed simulations of the streaking process and its application in real-time measurements of nanoscale near-fields are required. Johannes Schötz performed such simulations and shows them in his thesis.

The thesis not only describes the first steps into the new territory of attosecond resolved measurements on nanostructures, but it is also written such that it provides guidance to a newcomer. The thesis of Johannes Schötz is of high relevance to future research in attosecond nanophysics and I wish that his ground breaking work will find the wide and interested readership that it certainly deserves.

Prof. Matthias Kling

Ultrafast Nanophotonics Group,
Laboratory of Attosecond Physics
Department of Physics, Ludwig-Maximilians-Universität München &
Max Planck Institute of Quantum Optics, Garching, Germany

The Laboratory of Attosecond Physics (LAP)

Figure 0.1: Generation of attosecond pulses at LAP (Thorsten Naeser, MPQ)

The Laboratory of Attosecond Physics (LAP) is a unique facility for research on ultrafast particle motions outside of the atomic core. LAP is a joint facility of the Max Planck Institute of Quantum Optics (MPQ) and the Ludwig-Maximilians-Universität (LMU) Munich. The LAP team includes 150 scientists and students. Many of them are organized within the DFG cluster of excellence Munich-Centre for Advanced Photonics (MAP). The scientists are mainly interested in the motion of electrons, which change their probability density in quantum mechanical steps within attoseconds. In order to record such motion, the physicists have developed light flashes that last only attoseconds in duration. One attosecond is a billionth of a billionth of a second.

Physicists led by Prof. Dr. Ferenc Krausz, the current leader of the LAP team, generated and measured such extremely short light flashes for the first time in 2001. Since then, impressive insight into the mostly unknown world of electron motion has been gained world-wide, where the real-time dynamics of these particles can be followed after light-induced excitation.

The LAP research group Ultrafast Nanophotonics, which also Johannes Schötz is part of, is led by Prof. Dr. Matthias Kling. The research group investigates how electrons in complex materials collectively behave under the influence of intense laser light. In particular the physicists are interested in the dynamics and the control of electrons in molecules and nanostructures. Such light-induced electron motion occurs, for example, in semiconductors and dielectrics within attoseconds. Technically, the research could advance light-controlled nanoelectronics. With light frequencies in the Petahertz regime (10^{15} Hz) ultrafast switching times of electronic circuits could be achieved. This would advance current electronics by many (up to about 5) orders of magnitude.

Thorsten Naeser

Internet:
www.attoworld.de
www.munich-photonics.de

Contents

List of Figures

Introduction

Conventional electronics has reached over the past few years a fundamental limit, which restricts the clock speed from significantly exceeding 3.5 GHz [1,2] (see Fig. 1.1). To overcome this limitation, the control of electronic motion in signal processing by light pulses has been proposed [3] and called lightwave electronics. The period of light lies on the order of 1 fs, which would allow clock speeds 5-6 orders of magnitude faster than current state-of-the-art conventional electronics. Indeed, recent experiments showed, that by using few-cycle laser pules electrical currents can reversibly be induced in dielectric materials [4,5], with switching times on the femto- to attosecond time-scale.

One fundamental challenge in this approach is the issue of interconnects between such switches, as those are required to be on the nanoscale for use in on-chip integrated circuits and need to support high switching speeds [2]. This could be solved by the use of plasmonic nanostructures which allow nanoconfinement of light well beyond the diffraction limit and support frequencies from the visible to the infrared [6]. The extraordinary optical properties of nanoplasmonic systems arise due to the coupling of the dynamics of light with the collective electron motion on the nanoscale [7]. This leads besides the confinement of optical energy to an enhancement of the electromagnetic field near such nanostructures. Both effects are used in a growing number of diverse applications including chemical and bio-sensing with greatly enhanced sensitivity, nanoscopy, i.e. optical microscopy well below the diffraction limit, enhanced solar energy conversion and thermal cancer treatment [8].

Metal nanotips are a nanoplasmonic model system due to their widespread use in various nanoplasmonic applications and the relative simplicity and good control of their production process [9]. The coupling of light onto nanotips by use of nanogratings and the controlled transformation of the travelling surface plasmon polariton have been demonstrated [10, 11]. Besides the number of applications initiated by this approach, e.g. [12–14], this proves that nanotips are also an ideal system to study the possibility and fundamental limitations of the control of coherent electron motion in nanoplasmonics.

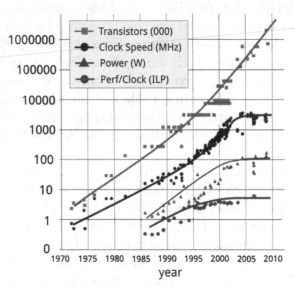

Figure 1.1: Intel CPU introductions. The clock speed is shown in dark circles. A stagnation at approximately 3.5 GHz beginning around 2005 is discernible (adapted from [1]).

Concerning the temporal evolution of the collective electron dynamics, first experiments on tungsten tips [15, 16] and gold tips [17] could proof the attosecond dynamics of electron emission from the apex of the tips when illuminated by few-cycle laserpulses. The reconstruction of this dynamics from the measurements however heavily relies on simulations. A technique which allows the measurement of electron dynamics on attosecond time scales directly in the temporal domain has become available about a decade ago with attosecond streaking metrology [18–20]. It is based on the process of high-harmonic generation with intense few-cycle IR laser pulses, which allows the production of isolated XUV-pulses with typical durations on the order of 100 as and energies around 100 eV. In the attosecond streaking pump-probe scheme, electrons are emitted by the XUV-pulse and subsequently accelerated by the strong few-cycle IR-pulse used for the production of the attosecond pulse. Originally used for the characterization of the IR- and XUV-pulse itself, it has been applied to study a number of processes in atoms [21] and from monocrystalline plane surfaces [22–24], which focus on the delay in photoemission delay of electrons from different quantum states. A number of recent theoretical studies suggest the feasibility of applying the concept of streaking to investigate the attosecond collective electron dynamics of nanoplasmonic systems [25–30].

In this thesis results of attosecond streaking on gold nanotips are presented. These are the first successful experiments of streaking measurements from nanosized solids. Numerical simulations where performed to understand the results of the experiment. Furthermore the effect of the neglect of transport effects in previous theoretical studies was examined. This thesis is organized as follows. In the second chapter an overview of the theoretical background necessary to describe attosecond streaking from nanoobjects is given. The third chapter describes the experimental techniques used for conducting attosecond streaking measurements. In the fourth chapter the influence of propagation effects of electrons inside the solid is examined in an semiclassical model and the limitations of those models are investigated. Finally in the fourth chapter the first experimental result on attosecond streaking from a metal nanotip are presented and compared to simulations. The experimental results in this thesis led to a paper which is waiting for publication [31].

2 Theoretical background

2.1 Ultrashort Laserpulses

The work treats the interaction of few-cycle laser pulses with matter. The basic description of the time evolution of the eletric field of a few-cycle laser pulse is most conveniently described using an envelope $f(t)$ and a phase function $(\phi(t))$. Due to its simplicity often a Gaussian envelope with linear phase is assumed:

$$E(t) = f(t) \cdot \cos(\phi(t)) = \exp(-2 \cdot ln(2)(t/\tau)^2) \cdot \cos(\omega_0 \cdot t + \phi_{CE}), \quad (2.1)$$

where ω_0 is the central frequency and τ is the intensity-FWHM. The Fourier transform leads to a connection between spectral width and temporal duration for a pulse with flat spectral phase [32]:

$$\tau_p \cdot \Delta\omega = 0.441 \cdot 2\pi. \quad (2.2)$$

where $\Delta\omega$ is the spectral intensity-FWHM. Few-cycle-pulses need an octave-spanning spectrum. As laser pulse described by Eq. 2.1 is plotted in Fig. 2.1. As an be seen for few-cycle laserpulses the CE-phase plays an important role in the time-evolution of the electric field.

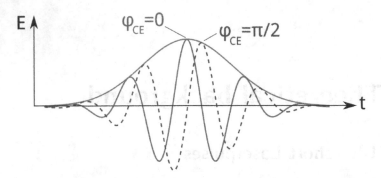

Figure 2.1: A 4.5 fs few-cycle laser pulse centered at 750 nm for two different carrier to envelope phases (ϕ_{CE} or CEP).

2.2 Maxwell's equations

The description of the interaction of electromagnetic fields with macroscopic matter is given in terms of Maxwell's equations [7]:

$$\nabla \cdot \vec{D} = \rho_{ext} \tag{2.3}$$

$$\nabla \cdot \vec{B} = \rho \tag{2.4}$$

$$\nabla \times \vec{E} = -\frac{\partial \vec{B}}{\partial t} \tag{2.5}$$

$$\nabla \times \vec{H} = \vec{J}_{ext} + \frac{\partial \vec{D}}{\partial t} \tag{2.6}$$

where $\vec{E}$ is the electric field, $\vec{D}$ is the electric displacement, $\vec{B}$ the magnetic field and $\vec{H}$ the auxiliary magnetic field and ρ and $\vec{J}$ are the external charge and current densities.

For nonmagnetic media $\vec{D}$ and $\vec{H}$ are linked to the electric and magnetic fields $\vec{E}$ and $\vec{B}$:

$$\vec{D} = \epsilon_0 \vec{E} + \vec{P} = \epsilon_0 \epsilon_r \vec{E} \tag{2.7}$$

$$\vec{B} = \mu_0 \vec{H}, \tag{2.8}$$

where ϵ_r is called the relative permittivity or dielectric function. Due to the interaction of the intrinsic charges and currents within the medium, the dielectric displacement is dependent on the response of the medium at surrounding positions and earlier times. This can be expressed as a temporal and spatial convolution:

$$\vec{D}(\vec{r}, t) = \epsilon_0 \int \mathrm{d}t' \int \mathrm{d}\vec{r}' \epsilon(\vec{r} - \vec{r}', t - t') \vec{E}(\vec{r}', t'). \qquad (2.9)$$

By switching to frequency-momentum-space via Fourier transform, we obtain:

$$\vec{E}(\vec{k}, \omega) = \epsilon_0 \epsilon_r(\vec{k}, \omega) \vec{E}(\vec{k}, \omega). \qquad (2.10)$$

For electromagnetic radiation the wavelength is usually large compared to the interaction length of the charges in the medium. It is thus save to approximate $\epsilon_r(\vec{k}, \omega) = \epsilon_r(\vec{k} = 0, \omega) = \epsilon_r(\omega)$, which is equivalent to assuming a local response. We will see in a later chapter, that for the problem of the passage of an electron through matter, the above assumption is not valid and the $\vec{k}$-dependence has to be kept. Generally, the finite response time of the medium, especially for ultrashort pulses, remains important and we have to keep the ω-dependence.

In the description of material properties of metals, the free-electron gas model has been very successful. By considering the polarization response of a free electron to an oscillating electric field $E(t) = E_0 \cdot e^{-i\omega t}$ the so-called Drude dielectric function can be derived:

$$\epsilon_D(\omega) = 1 - \frac{\omega_p^2}{\omega^2 + i\gamma\omega} \qquad (2.11)$$

where γ is the damping constant of electron motion. The plasma frequency ω_p is given by:

$$\omega_p = \sqrt{\frac{e^2 \cdot n}{\epsilon_0 m}}, \qquad (2.12)$$

where e is the electron charge, n the electron density and m the (effective) mass of an electron. Neglecting the damping, for $\omega = \omega_p$ the dielectric function vanishes and the electric field is given by an longitudinal depolarization field $\vec{E} = \frac{P}{\epsilon_0}$. This can be interpreted as an collective electron oscillation and is referred to as bulk plasmon [7]. It will play an important role when considering the energy loss of electrons passing through matter.

Mathematically finding the response of a nanoobject to an external field is a boundary value problem. Maxwell's equation lead to the following boundary conditions between medium 1 and medium 2 in the absence of free surface currents:

$$(\vec{E_1} - \vec{E_2}) \times \vec{n} = 0 \qquad (2.13)$$

$$(\vec{H_1} - \vec{H_2}) \times \vec{n} = 0 \qquad (2.14)$$

where E_1 (H_1) and E_2 (H_2) are the electric (auxiliary magnetic) fields on the left and right of the boundary respectively and $\vec{n}$ is the surface normal. In macroscopic Maxwell equation the surface charge is confined to an infinitesimally thin layer. In realistic materials the interaction of this charges leads to smearing of the surface charge on the length scale of the screening length [6]. The description of this effect requires nonlocal models and is a hot topic in the current research of nanoplasmonics [33, 34].

2.3 Nanoplasmonics

Nanoplasmonics allows overcoming the diffraction limit of light, by coupling collective electron dynamics (plasmons) to the oscillations of the electromagnetic field of light. [8]. Due to the coupling these excitations are called plasmon-polaritons [6]. Within the framework of second quantization of quantum mechanics they can be assigned a quasi-particle character. For classical incoming light (always the case in this thesis) they can however fully be described within the framework of Maxwell's equations [7]. The confinement of light is usually accompanied by enhancement of the electric field near the surface of the plasmonic object. Plasmon-polaritons are distinguished into surface plasmon-polaritons (SPP) and localized surface plasmons (-polaritons, LSP). Surface plasmon-polaritons are electromagnetic

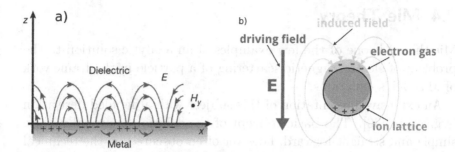

Figure 2.2: a): Illustration of a travelling surface plasmon polariton. The electric field is localized perpendicular to the surface. (from [35]) b): Localized surface plasmon polariton. The driving field causes oscillations of the electrons with respect to the lattice. Superposition of driving field and induced field leads to field enhancement at the poles.

waves which travel along a surface [7] and which are confined to the nanoscale perpendicular to the surface. They are simply solutions of Maxwell's equations coupling quasi-plane waves from both sides of the surface through Eq. 2.13. These solutions can be interpreted as SPPs if they are confined to the surface, which depends on the refractive index of the two media [7]. This is schematically in Fig. 2.2 a). SPPs are not restricted to plane surfaces and occur also on e.g. cylindrical or conical surfaces like the shank of a nanotip [10].

Localized surface plasmons by contrast occur when the dimensions of the particle fall below the wavelength of the incident light. The oscillating electric field of the incoming light causes a collective oscillation of the electrons with respect to the ion lattice [6], which leads to a confinement of optical energy on the order of the geometric features of the nanoobject and to field enhancement. This is schematically shown in Fig. 2.2 b).

2.4 Mie Theory

Mie theory is one of the few examples of an analytic solution to the problem of electromagnetic scattering of a particle in the framework of Maxwell's equations.

An extensive presentation of this subject with a detailed derivation is given in [36]. The basic concept of the solution is actually quite simple and straight-forward, however often obscured by the technical difficulties encountered on the way. We will therefore only outline the general formalism, which is independent of the geometry before briefly sketching the explicit solution for a sphere. The basic formalism starts from the Maxwell equations in a linear homogeneous isotropic medium in frequency-space:

$$\nabla \times \vec{E} = i\omega\mu\vec{H} \tag{2.15}$$

$$\nabla \times \vec{H} = -i\omega\epsilon\vec{E} \tag{2.16}$$

$$\nabla \cdot \vec{E} = 0 \tag{2.17}$$

$$\nabla \cdot \vec{H} = 0. \tag{2.18}$$

From this the well known wave equations can be derived:

$$\Delta\vec{E} + k^2\vec{E} = 0, \tag{2.19}$$

where $k^2 = \omega^2\epsilon\mu$ and analogously for $\vec{H}$. The first decisive step in solving such an equation is by making a general ansatz:

$$\vec{M} = \nabla \times \left(\vec{v}\psi\right) = -\vec{v} \times \left(\nabla\psi\right) \tag{2.20}$$

where $\vec{v}$ is a vector and ψ a scalar function. Plugging this ansatz into the wave-equation leads to

$$\Delta\vec{M} + k^2\vec{M} = \nabla \times \left(\Delta\psi + k^2\psi\right) \tag{2.21}$$

That means for $\vec{M}$ to satisfy the wave-equation, we only have to solve the scalar wave-equation for ψ. So far, we have not said anything about $\vec{v}$. It can be shown, that for Eq. 2.21 to hold, $\vec{v}$ can either be a (arbitrary) constant vector or the position vector $\vec{r}$ [37]. Moreover we notice that another independent solution to Eq. 2.19 is given by

$$, \vec{N} = \frac{1}{k} \nabla \times \vec{M} \qquad (2.22)$$

and that $\nabla \times \vec{N} = k\vec{M}$, using identities for the curl-operator. We can now write the solutions for $\vec{E}$ and $\vec{H}$ in terms of $\vec{M}$ and $\vec{N}$, the electric and auxiliary magnetic fields are linked by 2.15.
$\vec{M}$ and $\vec{N}$ are called vector harmonics and ψ the generating function. The two independent vector harmonics correspond to the two polarizations of light in a homogeneous medium. [37].

In order to solve the actual scattering problem, we first have to solve the scalar wave-equation to obtain a (possibly complete) set of solutions ψ. Then, the general fields inside and outside of the scatterer are expressed through the vector harmonics. Finally, the scattering solution is obtained by imposing the boundary conditions on the parallel components of $\vec{E}$ and $\vec{H}$.

The choice of the coordinate system clearly depends on the symmetry of the problem. Only for three geometries, namely spherical, spheroidal and cylindrical, exact analytic solutions are known [37]. We will briefly sketch the solution for light scattering of a sphere.

The scalar wave equation in a spherical coordinate system reads:

$$\frac{1}{r^2}\frac{\partial}{\partial r}\left(r^2\frac{\partial\psi}{\partial r}\right)+\frac{1}{r^2\sin\theta}\frac{\partial}{\partial\theta}\left(\sin\theta\frac{\partial\psi}{\partial\theta}\right)+\frac{1}{r^2\sin\theta}\frac{\partial^2\psi}{\partial\phi^2}+k^2\psi = 0. \quad (2.23)$$

By making a product ansatz $\psi = R(r)\Theta(\theta)\Phi(\phi)$ we get the set of even (subscript e) and odd (subscript o) solutions:

$$\psi_{emn} = \cos(m\phi)P_n^m(\cos\theta)z_n(kr) \qquad (2.24)$$

$$\psi_{omn} = \sin(m\phi)P_n^m(\cos\theta)z_n(kr), \qquad (2.25)$$

where P_n^m are the associated Legendre Polynomials, and z_n stands for any pair of the spherical Bessel functions of the first kind j_n, the second kind y_n, or the spherical Hankel functions of the first and second kind $h_n^{(1)}$ and $h_n^{(2)}$. The set of functions ψ_{emn} and ψ_{omn} forms a complete set. The general electric field can now be written as:

$$\vec{E} = \sum B_{emn}\vec{M}_{emn} + B_{omn}\vec{M}_{omn} + A_{emn}\vec{N}_{emn} + A_{omn}\vec{N}_{omn}. \quad (2.26)$$

The fields outside the sphere a expressed as a sum of incident E_i and scattered field E_s. Using the orthogonality of the vector spherical harmonics, in principle any incident field can be expanded as such a sum. For a plane wave one obtains:

$$E_i = E_0 exp(-ikz) \cdot \vec{e}_x = E_0 exp(-ikr\cos\theta) \cdot \vec{e}_x =$$

$$= E_0 \sum i^n \frac{2n+1}{n(n+1)} \left(\vec{M}_{o1n}^{(1)} - i\vec{N}_{e1n}^{(1)} \right) \quad (2.27)$$

where the superscript (1) indicates the use of the Bessel function of the first kind. Due to the orthogonality of the vector spherical harmonics we only need to consider the terms $\vec{M}_{o1n}$ and $\vec{N}_{e1n}$ for the scattered fields and the fields inside the sphere. Taking the boundary condition of vanishing scattered fields at infinity, and finite fields at the origin we obtain as ansatz for the field inside the sphere E_1:

$$E_1 = \sum E_n \left(c_n \vec{M}_{o1n}^{(1)} - id_n \vec{N}_{e1n}^{(1)} \right) \quad (2.28)$$

and for the scattered field E_s:

$$E_s = \sum ia_n \vec{N}_{e1n}^{(3)} - b_n \vec{M}_{o1n}^{(3)}, \quad (2.29)$$

where the superscript (3) denotes the usage of the Hankel function of the first kind. The coefficients a_n, b_n, c_n and d_n now have to be determined by imposing the boundary conditions on $\vec{E}$ and $\vec{H}$ (which follow from 2.15):

$$\left(E_i + E_s - E_1 \right) \times \vec{e}_r|_{r=R} = 0 \quad (2.30)$$

$$\left(H_i + H_s - H_1 \right) \times \vec{e}_r|_{r=R} = 0. \quad (2.31)$$

Using the size parameter $x = kR$ and the relative refractive index $m = k_1/k$ one obtains the solution:

$$c_n = \frac{\mu_1 j_n(x)[xh_n^{(1)}(x)]' - \mu_1 h_n^{(1)}(x)[xj_n(x)]'}{\mu_1 j_n(x)[xh_n^{(1)}(x)]' - \mu h_n^{(1)}(x)[mxj_n(mx)]'} \tag{2.32}$$

$$d_n = \frac{\mu_1 m j_n(x)[xh_n^{(1)}(x)]' - \mu_1 m h_n^{(1)}(x)[xj_n(x)]'}{\mu m^2 j_n(mx)[xh_n^{(1)}(x)]' - \mu_1 h_n^{(1)}(x)[mxj_n(mx)]'} \tag{2.33}$$

$$a_n = \frac{\mu m^2 j_n(x)[xj_n(x)]' - \mu_1 j_n^{(1)}(x)[mxj_n(mx)]'}{\mu m^2 j_n(mx)[xh_n^{(1)}(x)]' - \mu_1 h_n^{(1)}(x)[mxj_n(mx)]'} \tag{2.34}$$

$$b_n = \frac{\mu_1 j_n(mx)[xj_n(x)]' - \mu j_n^{(1)}(x)[mxj_n(mx)]'}{\mu_1 j_n(mx)[xh_n^{(1)}(x)]' - \mu h_n^{(1)}(x)[mxj_n(mx)]'}. \tag{2.35}$$

The importance of Mie theory in the field of nanoscience is manifold. Besides light scattering, it also allows the description of the plasmon modes and near-fields of a sphere and traveling plasmon modes of a cylinder. The general formalism can be used to obtain for example the fiber modes encountered in Super-Continuum-Generation in hollow-core-fibers (see section 3.1). Last but not least it is often used as a benchmark for computational algorithms.

2.5 Attosecond streaking

2.5.1 Fundamentals of attosecond streaking

The basic principle of attosecond streaking can be understood in a classical picture considering the electron as a point particle. Attosecond streaking is a pump-probe scheme with an XUV attosecond pump pulse and a strong few-cycle IR pump pulse. The XUV pulse leads to photoemission and the emitted electrons are subsequently accelerated by the oscillating field of the IR pulse . This is schematically shown in Fig. 2.3 a). The initial velocity of the electrons depends on the XUV energy which is on the order of 100 eV. In this non-relativistic

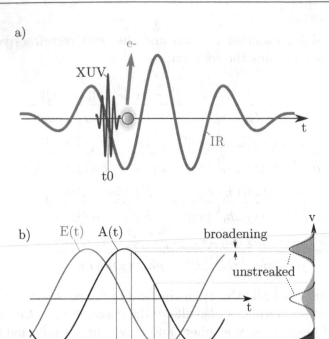

Figure 2.3: Principle of attosecond streaking: a) The time t_0 of electron emission depends on the timing of the XUV attosecond
pulse and the few cycle IR pulse. b) The relation between
the emission time and the velocity shift experienced by the
photoemitted electrons. Features of the XUV pulse are encoded in the change of the shape of the electron spectrum,
e.g. the finite temporal duration leads to a delay dependent
broadening of the spectrum compared to the unstreaked case
(thick dashed line).

regime, omitting the magnetic field, the change of the electron velocity emitted in polarization direction can be written as:

$$\Delta v = -\frac{e}{m} \int_{t_0}^{\infty} dt\, E(t), \tag{2.36}$$

where it has been assumed that the electric field is homogeneous in space, which is usually a good assumption since the XUV focus is much smaller than the IR focus [38]. Using the relation of the electric field to the vector potential $\frac{\partial \vec{A}}{\partial t} = -E(t)$ and the vanishing DC-component of a laser pulse $\int_{-\infty}^{\infty} E(t) = \mathcal{F}[E(\omega = 0)] = 0 = A(\infty)$, the above equation can be rewritten as:

$$\Delta v = -\frac{e}{m} A(t). \tag{2.37}$$

Depending on the time of emission the electron bunch initiated by the attosecond XUV pulse experiences a velocity shift proportional to the vector potential. This is schematically depicted in Fig 2.3 b).

The final energy is given by

$$E(t_0) = \frac{m}{2}(v_0 + \Delta v)^2 =$$

$$= (\hbar\omega - I_p) - \sqrt{\frac{2}{m}(\hbar\omega - I_p)}\frac{e}{m} \cdot A(t_0) + \left(\frac{e}{m}A(t_0)\right)^2, \tag{2.38}$$

where the last term is usually negligible even for relatively high amplitudes [38]. While the overall shift reflects the vector potential of the IR pulse, the change of the shape of the electron pulse reflects properties of the exciting XUV pulse. In Fig. 2.3 a) this is schematically shown in the delay dependent broadening of the velocity distribution due to the finite temporal duration of the XUV pulse. Also higher order features of the XUV pulse like chirp are encoded in the spectrum and using attosecond streaking both pulses can be characterized [38]. An experimental streaking trace from Neon gas is shown in Fig. 2.4 Attosecond streaking enables to resolve processes on the order of the pulse length of the XUV pulse.

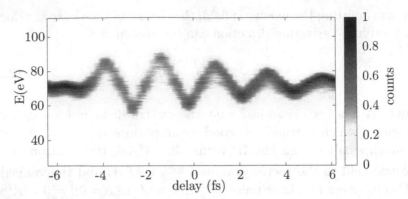

Figure 2.4: An experimental attosecond streaking trace from Neon measured with our setup.

2.5.2 Attosecond streaking from solids

Attosecond streaking has been applied to measure processes on monocrystalline surfaces. Since attosecond streaking is a time-resolved photoemission process, first the theoretical description of photoemission without the IR probe is considered before a brief overview of the existing models of attosecond streaking from solids is given.

Photoemission

Historically one distinguishes between the one-step and three-step picture [39]. The photoemission probability in the one-step pictures and the first step in the three-step picture for weak exciting light fields is given in terms of Fermi's golden rule:

$$P(i \to f) \propto |<\psi_f|\mathcal{H}_{\text{PE}}|\psi_i>|^2 \delta(\hbar\omega - E_f + E_i), \qquad (2.39)$$

where ψ_i and ψ_f are the initial and final many-body states respectively, $\hbar\omega$ is the energy of the exciting light. $\mathcal{H}_{\mathcal{PE}}$ is the photoemission operator, which is given by [40]:

$$\mathcal{H}_{\mathcal{PE}} = \frac{ie\hbar}{2m}\left(\nabla \cdot \vec{A} + \vec{A} \cdot \nabla\right) + \frac{e^2}{m}\vec{A^2}, \qquad (2.40)$$

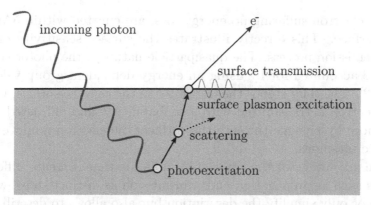

Figure 2.5: The photoemission process in the three step picture.

where $\vec{A}$ is the vector potential.

State-of-the-art models use density functional theory (DFT) to calculate the inital state and are thus able to consistently incorporate the crystal symmetry and effects such as surface reconstruction and recombination [39]. The models differ in the treatment of the final state and electron scattering within the solid. In the one-step model the final state is generally computed by Quantum Field theoretical methods as a state which behaves asymptotically as a free electron and which is scattered inside the solid. It coherently takes into account effects like intrinsic and extrinsic excitations at the price of high computational demand.

In the three-step model, the final state is an eigenstate of the crystal and the computation of Eq. 2.39 leads to the probability of the first step. The three steps are schematically depicted in Fig. 2.5. The second step consists in the transport of the electron to the surface. The probability $P(s)$ that the electron undergoes a scattering event while travelling a distance s, is defined by the material specific mean-free path λ:

$$P(s) \propto e^{-s/\lambda}. \tag{2.41}$$

The inelastic mean free path of most metals for electron energies around 100eV lies in the region of 5 Å [41]. That means that 63%

of the electron suffering no energy loss, are emitted within 5Åfrom the surface. This directly illustrates the surface sensitivity of the photoemission process. The quasiparticle nature of the photoelectron can be accounted for by using an energy dependent group velocity v_G. [42]. The third step is the transmission through the surface, where the electron is transmitted with a probability T and diffracted. Furthermore by passage through the surface, the electron might excite surface plasmons.

The advantage of the three-step model is that it treats different effects such as propagation and transmission as distinct steps, which does not only simplify the description but also allows to describe the the different effects with different degrees of accuracy.

To get a description of attosecond streaking based on this models, the effect of the IR pulse has to be incorporated. In the one-step picture this leads to a different final state, which is again difficult to calculate, whereas in the three-step model it can simply be incorporated into the propagation of the electron.

Attosecond Streaking from Plane Surfaces

The first experiments of attosecond streaking from a plane tungsten surface [22], and subsequent experiments on rhenium [23] and magnesium [24], which focused on measuring time delays between valence and core bands, initiated a number of theoretical models. A recent overview models is given in [33]. All but one [42] use a quantum mechanical description in the single-active electron approximation. The IR-streaking field can either be taken into account by directly solving the time-dependent Schrdinger equation or by employing damped Volkov-states as final states in Eq. 2.39 [33]. Using the symmetry of the experiments, the quasi-perpendicular polarization of XUV and IR with respect to the surface and normal emission, allows a considerable simplification of the models. Different factors such as dispersion or different localization and energy-dependent scattering where used to explain the observed time-shifts of around 100as for tungsten and rhenium and 0 as for magnesium.

Attosecond streaking from nanoplasmonic objets

A number of theoretical studies [25–30] have suggested the use of attosecond streaking for the characterization of nanoplasmonic near-fields. Several points restrict the models for plane surfaces to be used for nanoplasmonics objects. First, due to the changed geometry, the symmetry is reduced, IR and XUV fields are generally not normal to the surface anymore and non-normal electron emission has to be considered. Additionally, the IR fields show pronounced inhomogeni-ties and the considered objects are usually polycristalline. Therefore employed models to describe the attosecond streaking process from nanoobjects are considerably simplified. Electrons are treated in a classical framework. Most models assume electron emission from a narrow band and neglect scattered electrons [25, 27–29] or use an experimental spectrum [26]. Furthermore emission from the surface is assumed and any effects of electron emission and propagation in the solid are neglected, except for [29].

With the first successful attosecond streaking experiments on nanos tructures (see Chapter 5), a more detailed model to study the influence of effects neglected above is necessary. Such a model has been devel-oped in this work in the framework of the three-step model. Although the model should be widely applicable, we limit our discussion to attosecond streaking from gold and XUV energies in the region from 80 to 110eV region, the conditions found in experiment. We use a Monte-Carlo algorithm. The XUV beam is described by an Gaussian-profile in space and time and propagates along a straight ray-like line. Over the distances relevant for photoemission the attenuation and the temporal distortion of the XUV-pulse is neglected. This is in agreement with Mie calculations on spheres and cylinders. Since the polarization of the XUV field is numerically hard to calculate and due to the unknown crystal symmetry the photoexcitation is assumed to be isotropic and homogenous and is restricted to a layer a few times the inelastic mean free path from the surface. The initial energy of the electrons is given by the convolution of the experimentally XUV-spectrum and the valence band density of states of gold calculated by

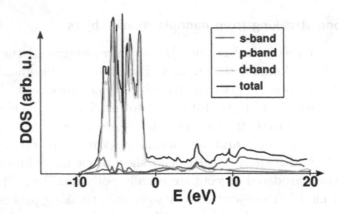

Figure 2.6: partial and total density of states of the valence band of gold
calculated by full-potential linear muffin-tin orbital (FP-
LMTO) implementation of density-functional theory (DFT)
in the local-density approximation (LDA) [43] [adapted from
[44]]

density functional theory (DFT) in the full-potential linear muffin-tin
implementation (FP-LMTO) using the local density approximation
(LDA) [43, 44]. The employed DOS of gold is shown in Fig. 2.6. The
electron is subsequently propagated through the medium subject to
elastic and inelastic scattering until it reaches the surface, where it is
diffracted and may lead to surface excitations. The relation between
the DOS and the resulting (unstreaked) photoelectron spectrum is
shown in Fig. 2.7. A free electron dispersion is assumed. Scattering
and surface transmission are described in great detail in the next chapter. All the time from its birth, the electron is subject to the IR field
and the following classical equation of motion is solved numerically:

$$\frac{\mathrm{d}\vec{v}}{\mathrm{d}t} = -\frac{e}{m}\vec{E}_{IR}(\vec{r}, t). \tag{2.42}$$

Electron scattering, which changes the electron energy and direction,
is assumed to occur instantly. The propagation is stopped when all
fields have substantially decayed.

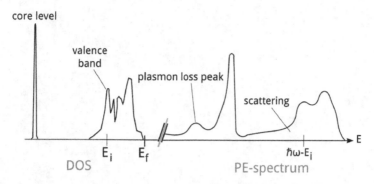

Figure 2.7: Illustration of the relation between the photoemission spectrum and the DOS. The finite width of the exciting photon spectrum as well as scattering lead to a broadening of the features in the photoelectron spectrum.

3 Experimental methods and setup

The attosecond streaking technique requires CEP-controlled high-power few-cycle laserpulses to produce isolated attosecond XUV-pulses. In this section a short overview of the setup for the production of few-cycle pulses is given before turning to the generation of isolated XUV pulses and the implementation of the attosecond streaking technique itself.

3.1 Generation of ultrashort laserpulses

The Ti:Sa-Laser is up-to-date the workhorse of attosecond physics because of the broad bandwidth which it supports and which allows the generation of few-cycle laserpulses from an oscillator. Fig. 3.1 shows a schematic overview of the lasersystem used in our experiments. The lasersource is as commercially available chirped-pulse amplifier (FEMTOPOWER Compact Pro). The first part is a Kerr-mode-locked Ti:Sa-oscillator which delivers 3.5 nJ pulses with a duration of 7 fs covering a spectral range from 620 nm to 1000 nm at a repetition rate of 70 MHz. It is equipped with a module for CEP-stabilization (see subsequent section). A pulse-picker sends pulses at a repetition rate of 1 kHz to the subsequent amplification stage. Then the pulses are sent through a SF-57 glass stretcher where they are elongated to around 10 ps by introducing negative chirp. For fine tuning of the dispersion a programmable acusto-optic dispersive filter is used (FASTLITE Dazzler) before finally passing the chirped pulses to the Ti:Sa 10-pass amplifier. After amplification the pulses have an energy of around 2 mJ and due to gain narrowing a spectral bandwidth

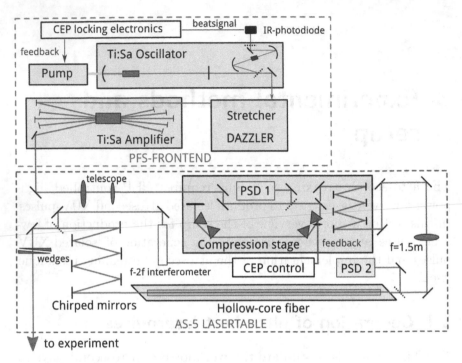

Figure 3.1: Schematic overview over the AS5-laser-system used for the
experiments. [Adapted from [45]]

of around 100 nm centered at 800 nm. They are then sent to the
AS5-lasertable, where they pass through a telescope before entering
the pulse compressor, which implements a hybrid approach. First
a prism-compressor is used to precompress the pulses. However to
avoid pulse distortions due to optical nonlinearities in the last prism
caused by high-intensities the pulses are not fully compressed. This is
achieved in second stage with a set of highly dispersive mirrors. After
this, the pulse is compressed to a duration of around 27 fs with an
energy of approximately 1.6 mJ. To avoid beam-pointing instabilities
a beam-stabilization system, which measures pointing fluctuations
via Position-Sensitive-Detectors (PSD), is employed before and after
the compressor. For spectral broadening the pulse is focused into a

hollow-core fiber with an inner diameter of 275 μm. The fiber is filled with neon gas at an optimized pressure between 2.6 and 3.2 bar. The spectral broadening occurs due to the optical Kerr-effect. It can be described by the second order nonlinear refractive index n_2 and leads to an intensity dependent refractive index :

$$n(t) = n_0 + n_2 \cdot I(t), \qquad (3.1)$$

where n_0 is the linear refractive index and $I(t)$ the intensity. This causes an additional time dependence of the instantaneous phase of the pulse $\phi(z,t) = -\omega \cdot t + n(t)k_z z$, which changes the effective frequency ω_{eff}:

$$\omega_{eff} = \frac{\partial}{\partial t}\phi(z,t) = \omega - n_2 z \frac{\partial I(z,t)}{\partial t}. \qquad (3.2)$$

The change of intensity within the pulse leads to the generation of additional red (blue) frequency components in the leading (trailing) part. Additionally so-called self-steepening occurs, which leads to an steeper trailing edge and therefore favours the generation of blue components [32].

The resulting positively chirped pulse is recompressed using chirped mirrors. The phase of the output pulse is however not trivial over the whole part of the generated spectrum and therefore not the whole bandwidth can be used effectively for pulse compression [38]. For fine-tuning and compensation of day-to-day variations a pair of glass wedges is used. A small portion of the beam is sent into another CEP-stabilization system. The rest of the beam is sent to the experiment.

3.2 CEP-stabilization

The generation of isolated attosecond pulses used in streaking spectroscopy heavily relies on waveforms with a controlled CEP. In our setup the CEP is controlled at two different positions namely directly at the oscillator and after the spectral broadening in the hollow-core-fiber and two different techniques are employed.

a)

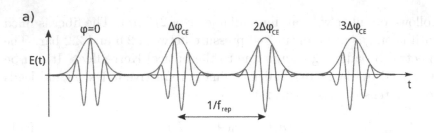

b)

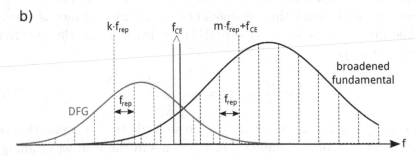

Figure 3.2: a): Pulses from a mode-locked oscillator with repetition rate
f_{rep} and constant phase slip $\Delta\phi_{CE}$. b) The phase slip leads
to an offset of the frequency comb of the fundamental laser
by f_{CE}. The frequency comb of the difference frequency
signal (DFG) lies at multiple integers of f_{rep}. Interference
of the fundamental and the DFG-signal leads to a beating
signal.

As schematically shown in Fig. 3.2 a), mode-locked oscillators
deliver pulses with a well defined pulse envelope and repetition rate,
however there is generally a pulse-to-pulse shift of the carrier-to-
envelope phase $\Delta\phi_{CE}$. In the frequency domain this corresponds
to a shift of the teeth of the frequency comb $f_n = n \cdot f_{rep}$ by the
so-called carrier-envelope phase offset frequency $f_{CE} = f_{rep}\Delta\phi_{CE}/2\pi$.
In our setup the offset-frequency is measured by focusing a part of
the oscillator output into a periodically-poled lithium niobate crystal
(PPLN). Via the second order nonlinearity a difference frequency

signal (DFG) is produced, the teeth of which lie at multiple integers of the repetition frequency:

$$f'_m = f_k - f_n = (k \cdot f_{rep} + f_{CE}) - (n \cdot f_{rep} + f_0) = m \cdot f_{rep}. \quad (3.3)$$

where $k - n = m$. This is schematically shown in Fig. 3.2 b). In the region where both spectra overlap a beating can be observed. The measurement of f_{CE} via the beating signal can be understood by considering the temporal intensity $I(t)$ on a fast photodiode, which is able to resolve frequencies on the order of f_{rep} and simply averages over the frequencies which occur in the comb:

$$I(t) \propto\, < \sum_{m,l} [cos(\omega_l \cdot t) + cos(\omega'_m \cdot t)]^2 > \quad (3.4)$$

$$=< \sum_{m,l} cos(\omega_l \cdot t)^2 + cos(\omega'_m \cdot t)^2 + 2 \cdot cos(\omega_l \cdot t) cos(\omega'_m \cdot t) >$$

$$= \sum_{m,l} [1+ < cos(\omega_l \cdot t + \omega'_m \cdot t) > + < cos(\omega_l \cdot t - \omega'_m \cdot t) >]$$

$$= \sum_{m,l} [\text{const.} + cos(\omega_l \cdot t - \omega'_m \cdot t)],$$

where the prime denotes the DFG-comb. Relative amplitudes of the different contributions have been neglected. The beating signal thus contains frequency components at $f_l - f'_m = f_{CE}$ for $l = m$ which can be determined by Fourier analysing the time signal of the photodiode. The rest of the spectrum, where no beating occurs is filtered out. By considering different pairings of frequencies in the above equation, it can be understood why f_{CE} is usually stabilized to $f_{rep}/4$. At integer values of f_{rep} it would overlap with contributions from $\{f_{l+n}, f_l\}$ and at $f_{rep}/2$ with the signal stemming from $\{f_{l-1}, f'_l\}$. In the setup f_{CE} is referenced to $f_{rep}/4$ in the locking electronics (MENLO systems) and stabilized via a change of the intracavity dispersion by modulating the pump power with an acusto-optical modulator (AOM). The pulse picker ensures that only pulses with identical CEP are amplified.

After the oscillator CEP-drifts might occur due to intensity fluctuations in the amplifier and pointing instabilities in the whole setup. Therefore the CEP of the recompressed pulse after the chirped mirrors is measured. This is done by focusing a part of the beam into a β-barium borate BBO-crystal which is optimized for SHG. The resulting spectrum $I(\omega)$ of the overlap of the fundamental and SHG is measured via a spectrometer and is given by [45]:

$$I(\omega) = I_f(\omega) + I_{2f}(\omega) + \sqrt{I_f(\omega)I_{2f}(\omega)}cos(\phi_f(\omega) - \phi_{2f}(\omega) + \omega\tau + \phi_{CE}).$$
$$(3.5)$$

Due to the delay τ of the spectral components of fundamental and SHG signal a modulation of the measured intensity occurs in the region of spectral overlap. This modulation is shifted ϕ_{CE}. By Fourier analysing the modulation, the relative CEP can be calculated [46]. By a feedback on one of the prisms in the hybrid compressor, the dispersion can be changed and the CEP stabilized.

We note, that in connection with the first technique one speaks of "f-to-0"-technique, and in connection with the second of "f-to-2f"-technique [41]. The term however only describes whether the the fundamental is referenced to the DFG or the SHG signal. Variation of the first technique, where the SHG-signal is used are widespread [38].

3.3 High-Harmonic Generation

High-harmonic generation (HHG) is the decisive process for the generation of attosecond XUV pulses and is achieved by focusing a high-intensity laser-beam into gas. Most of the features of HHG can be understood in a simple semi-classical model, the so-called 'simple-man's model' [38], which can be decomposed into three steps. A schematic illustration of this model is shown in Fig. 3.3 a). First, at the crest of the electric field, the incident laser causes a strong distortion of the Coulomb potential of the gas atom, which leads to ionization by electron tunnelling through the potential barrier. The

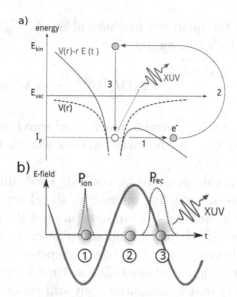

Figure 3.3: a) The simple-man's model of High-Harmonic Generation. The dashed line shows the unperturbed Coulomb potential, the red line the total potential at the time of electron emission. b) The timing with respect to the driving laser. The dashed curves schematically show the ionization (P_{ion}) and recollision probabilities (P_{rec}).

emitted electrons, initially at rest, are accelerated away from the parent ion. Secondly, due to its temporal evolution, the field will reverse its direction. The electrons are decelerated and will finally travel back to the parent ion. Thirdly upon recollision and recombination with the parent ion, a photon with an energy equal to the sum of kinetic energy and ionization potential I_p is created. The maximum photon energy E_{cutoff} predicted by this model is given by [38]:

$$E_{cutoff} = 3.17 \cdot U_p + I_p, \qquad (3.6)$$

which is close to the quantum mechanical result. U_p is the pondero-motive energy which is given by:

$$U_p = \frac{(eE_0(\tau))^2}{4m\omega_0^2} \approx 9.33 I_L[10^{14}\frac{W}{cm^2}] \cdot \lambda_0[\mu m]^2, \tag{3.7}$$

where ω_0 and λ_0 are the central frequency and wavelength respectively. $E_0(\tau)$ is the amplitude of the instantaneous electric field oscillation and I_L is the intensity.

Two important things are to be noticed. First, due to the highly nonlinear process of tunneling ionization, the electrons are emitted only during a well confined time interval around the field maximum. As a consequence also the recollision and photon emission is synchronized with the driving laser. Due to the dispersion of the electrons while travelling through the laser-field, the time interval of recollision is broadened compared to ionization, but still confined to a fraction of the optical period. This is schematically shown in Fig. 3.3 b). The process of ionization and recollision repeats every half cycle and leads to the production of attosecond pulse trains.

Secondly, for few-cycle laser-pulses the maximum electric field amplitude depends on the CE-phase (see Fig. 2.1) and with that through Eq. 3.6 the maximum photon energy produced in HHG. For $\phi_{CE} = 0$, the highest photon energies are produced only during the central half-cycle. By spectrally filtering out the highest energy photons, which are only produced during the central half-cycle, isolated attosecond pulses can be generated. This is schematically shown in Fig. 3.4. The basic characteristics of HHG have so far been discussed in the intuitive semi-classical picture. A quantum mechanical description becomes necessary for example when examining the efficiency of the HHG process or when considering phase-matching of the HHG-beam [38]. In our setup the laser is focused with a peak intensity of around $5 \cdot 10^{14} \frac{W}{cm^2}$ into a neon gas jet. By use of Eq. 3.6 this leads to a cutoff-energy of approximately 105 eV. Fig. 3.5 shows the attosecond pulses produced in our experiments after spectral filtering and dispersion compensation as measured by attosecond streaking from neon gas.

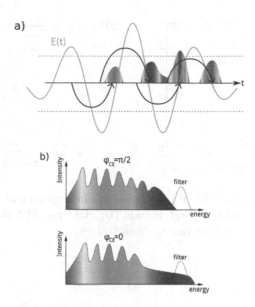

Figure 3.4: a): The HHG-process for $\phi_{CE} = 0$. The color encodes the energy of the produced XUV-photons. The highest energies are only produced during the central field oscillation. b): Sketch of the HHG-spectrum for two different CE-phases. The spetral filter, which allows generation of isolated attosecond pulses, is indicated as a black dashed line. [adapted and corrected from [45]].

They lie around 95 eV with a bandwidth of around 8 eV given by the spectral filter and with a duration of approximately 220 as.

3.4 AS5-Beamline

Fig. 3.6 shows a schematic overview of the AS5-attosecond beamline used in the experiments. Because the XUV pulses get easily absorbed the whole setup resides in vacuum chambers, evacuated to pressures around 10^{-5} mbar or lower. The few-cycle IR-beam coming from the lasertable is sent into the HHG-chamber where it is focused into a neon-gas target by a mirror with a focal length of 50 cm. After

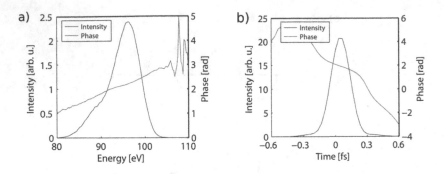

Figure 3.5: A typical XUV pulse used in the experiment in the spectral
(a) and temporal domain (b), determined from an experimental streaking trace. [from [45].

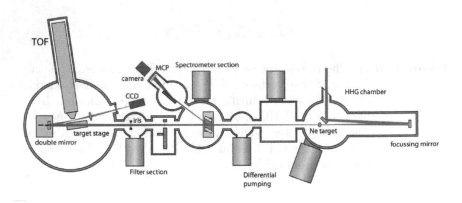

Figure 3.6: Overview over the AS5-attosecond beamline [45].

passing through a system of differential pumping stages, the combined
HHG and IR beam reaches the XUV-characterization stage. There,
a gold mirror can be driven into the beam path, which projects the
XUV onto a grating spectrometer, consisting of a grazing incidence
grating and a MCP with a fluorescent screen. The beam then passes
through the filter section, where XUV and IR are spatially separated.
The outer beam passes through a pellicle, blocking the XUV, and
the inner part passes through a Zirconium foil, blocking the IR. The

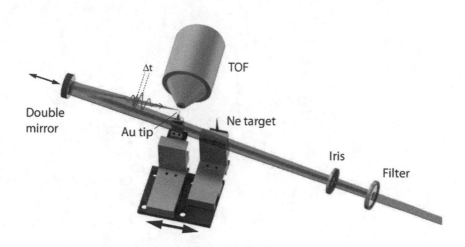

Figure 3.7: Detailed view on the TOF, double mirror and target stages
[45].

beam is then sent into the experimental chamber onto a spherical
double mirror. The position of the inner part of the mirror is controled
via a high-precision piezo-stage, which allows to introduce the delay
between XUV and IR necessary for the attosecond streaking pump-
probe scheme. It has a focal length of 12.5 cm and focuses both beams
under a slight angle onto the targets mounted on the target stage in
front of the time-of-flight spectrometer (Käasdorf TOF ETF-20). This
is shown in more detail in Fig. 3.7. The spectrometer registers the
arrival of electrons via an MCP which produces a fast voltage signal.
The timing of this signal is then recorded with respect to a trigger
signal, produced by the previous IR pulse, by a fast multiple-event
time digitizer card (TDC FASTCOMTec p7889). From this the time-
of-flight spectrum of the electrons can be deduced and converted to
energy with the use of calibration curves provided by the manufacturer
of the TOF.

4 Electron scattering in solids

Electron scattering in solids is a problem encountered in many experiments for example Reflection Electron Energy Loss Spectroscopy (REELS), Low Energy Electron Diffraction (LEED), Transmission Electron Microscopy (TEM) and also in radiation therapy, radiation-protection and for radiation detectors.

In Scattering theory generally two types of scattering are distinguished, namely elastic and inelastic scattering. In inelastic scattering events, part of the energy of the projectile is used to promote the target to a different internal state, while in elastic scattering it remains the same.

The passage of electrons through solids and surfaces is in principle a complex many-body problem which is very difficult, if not impossible, to solve correctly. Therefore simplified models have to be used for their description. These models turn out quite differently for elastic and inelastic scattering, as will be elaborated in the present chapter.

4.1 Elastic Scattering

The elastic scattering of an electron with kinetic energy E is usually described as the interaction with a single free target atom, modelled via the spherically symmetric potential $V(r)$. The neglect of the crystal structure and bonds of the solid seems to be a severe approximation at first, however, as will be discussed below, it is justified for most cases.

Due to the spherical symmetry of the problem, the well established partial wave methods can be used. Although it can be found in any good textbook on Advanced Quantum Mechanics (e.g. [47]) we will shortly describe the general procedure here. Due to the symmetry the solution of the Schrödinger Equation (SE) can be expanded into different orbital angular momentum states with quantum number l. This leads to a separate one-dimensional equation for each l in the radial variable r. For large r, where the potential for a neutral atom is zero, the general solution is known analytically (it has the same form as the one encountered in Mie-theory [36]). The ansatz for the solution is a distorted plane wave, an incoming plane wave plus an outgoing spherical wave. By numerically integrating the SE for each angular momentum state l up to the radius where the potential vanishes (has reached it's asymptotic value) and matching it to the distorted plane wave solution, the differential cross section can easily be calculated.

In order to allow for a more accurate description in the framework of relativistic quantum mechanics, the Schrödinger equation has to be replaced by the Dirac equation, which also naturally accounts for the spin of the electron. The general procedure however remains the same.

There exist a number of databases for elastic scattering cross sections calculated by the above procedure, whereas experimental data exists only for a number of elements at specific energies. An overview and comparison of the databases is given in [48]. The main differences are the employed potentials, the covered energy range, the available elements and the relativistic/non-relativistic treatment. An alternative to databases is to use available codes, which calculate the desired quantities. In this thesis we will utilize the code-system ELSEPA [49], which has been used in a number of recent Monte-Carlo simulation papers involving electron transport, including a recent paper in the context of attosecond physics on metal nanotips [50] and streaking from tungsten surfaces [42]. It uses Dirac-Partial-Wave-Analysis and employs the state-of-the-art model, which allows the description of relatively low energy elastic scattering. It also allows the user to vary certain parameters in the potentials to study the effect on the scattering characteristics.

In the ELSEPA code-system the interaction is modelled using an optical model potential [51]:

$$V(r) = V_{st}(r) + V_{ex}(r) + V_{cp}(r) - i \cdot W_{abs}(r). \qquad (4.1)$$

where V_{st} is the electrostatic interaction. In ELSEPA it is obtained from Dirac-Hartree-Fock self-consistent density-functional calculations for a single atom.

V_{ex} is the exchange potential. It arises due to the Pauli-exclusion principle, the exchange antisymmetry for electrons. This is a non-local effect, which is difficult to handle in calculations. Therefore $V_{ex}(r)$ is derived from the local-density-approximation (LDA) of a homogeneous electron gas (HEG).

The terms V_{st} and V_{ex} are used in other calculations too and correspond to a first order Born-approximation [51], which is asymptotically exact for high energy electrons. For lower energies the polarization and the absorption of the atom play a role. V_{cp} is the correlation-polarization potential which accounts for the polarization via the classical long range polarization potential and the electron screening via a short range correlation potential obtained from LDA. W_{abs} is the absorptive potential. It is also derived from the electron energy loss through e-h-creation in a HEG.

Furthermore to allow for the description of elastic scattering within solids, the atomic V_{st} can be replaced by a muffin-tin potential. This is a common approach to describe electrons in solids. In the model used in ELSEPA it is assumed that all atomic electrons are confined to within a sphere of radius R_{mt}.

Fig. 4.1 b) shows the differential elastic cross section (DCS) for a gold atom for different energies. It is calculated using the optical model potential described above and a muffin-tin radius $R_{mt} = 1.37$ Å, the order of magnitude of which can be estimated from the lattice constant of gold $a_{lattice} = 4.11$ Å. For all energies the DCS is dominated by a strong forward scattering peak, which becomes more pronounced with increasing energy. In addition there appear minima in the DCS, which arise due to the interference of different angular

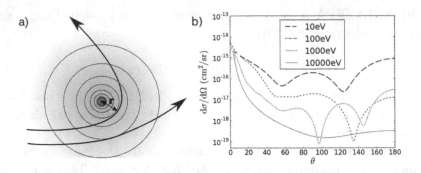

Figure 4.1: a) Classical picture of the relation between radial distance
and scattering angle. b) Differential cross section for gold for
different energies computed with a muffin-tin radius of 2.4
a_0. Note the logarithmic scale.

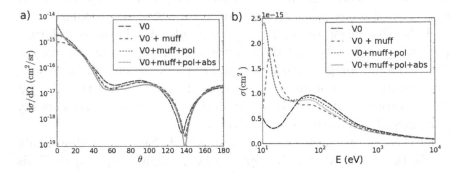

Figure 4.2: Effect of the muffin-tin model and the different higher order
terms on a) the differential cross section for gold at $93eV$
and b) on the total elastic cross section versus energy.

momentum contributions. The position, width and depth vary with
energy and eventually disappear for high energies.

To see the effect of different higher order contributions in the optical model potential a set of calculations were performed starting from
the combined electrostatic-exchange potential $V_{st} + V_{ex}$ and consecu-
tively switching on the muffin-tin option, the correlation-polarization

potential and finally the absorptive potential. As can be seen in Fig. 4.2 a) the muffin-tin option (green/wide dashes) slightly shifts the minima but more importantly lowers the DCS at small scattering angles. The reason for the latter is that it lowers the electron density and therefore the potential at higher radii compared to the atomic case. This can be understood already in a classical picture as illustrated in Fig. 4.1 a). The main effect of V_{cp} (red/narrow dashes) is to introduce a long-range polarization potential, which leads to a increase in $V(r)$ for high r. This in contrast, leads to an increase in the DCS at relatively low scattering angles. The influence on the distribution for higher scattering angles is minor. W_{abs} (light blue/solid line) lowers the DCS at all angles without drastically influencing the shape. The effect of the higher order corrections becomes less important for higher electron energies and the DCS practically converge. For lower energies however, the impact of V_{cp} and the muffin-tin option increases and they eventually dominate the distribution for energies below about 40 eV.

This is also illustrated in the total elastic cross section, as shown in Fig. 4.2b). While at high energies the different models practically deliver the same σ_{tot}. For 100 eV the differences are still only a few percent, whereas below 40 eV they might get as high as several 100%. In this region the biggest impact is from the muffin-tin model. This can also be seen in Fig. 4.3, where the energy dependence of the total elastic cross section for different muffin-tin radii is shown. The height and position of the maximum below 40 eV critically depends on R_{mt}. This can be understood by considering the de-Broglie wavelength λ_{DB} of the electron, which reaches in this energy region the size of the muffin-tin potential. The sensitivity on R_{mt} thus indicates the occurrence of a resonance.

There are a few points to conclude from the above analysis. First of all, the differential cross section for gold has a rather complicated structure in the energy region around $100eV$. Even for low energies it can not be approximated by an isotropic distribution. That implies that for Monte-Carlo simulations the calculated differential cross section should be used.

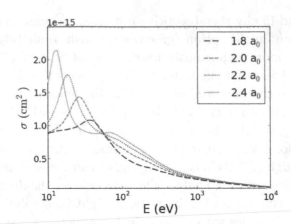

Figure 4.3: The dependence of the total elastic scattering cross section on the muffin-tin radius R_{mt} in units of the Bohr-radius a_0. The shift of the low energy maxima with the radius indicates the occurrence of a resonance.

Secondly, in this model for energies below 40eV the higher order corrections become dominant for elastic scattering. The occurrence of resonances in this energy region, lead to a drop of the elastic mean free path below the interatomic distance. This violates the assumption of a incoming plane wave for consecutive scattering event and leads to a break down of the model. A study for silica showed, by describing the electrons as Bloch-electrons and considering the different inelastic and quasi-elastic channels separately, that the elastic cross sections in the low-energy region are indeed significantly lower than calculated from models similiar to one the above [52].

Since for real solids we expect foremost that the potential at high r is different from the atomic case, we can conclude that for higher energies only the small scattering angles are considerably affected, as seen above. Small angle scattering below 10° only slightly affects the (incoherent) particle transport over small distances, which is the reason why atomic cross sections have been successfully applied for electron elastic scattering at solid surfaces [48].

There a two further points in the model which have to be discussed when applying the above results to photo-emission from solids. First, in real situations the incoming electron will not be a plane wave but more or less spatially confined wave-packet. However, if we assume that the wave-front is approximately plane over the length scale of the atomic radius, we can apply the above results. Furthermore, for low energies the actual band structure of the solid might strongly influence the dispersion relation of electron and the approximation of an electron as a free electron might be invalid.

Finally the symmetry and periodicity inside a crystal can lead to the well known Bragg diffractions in coherent scattering. However for this the electron wave-packet must be describable as a plane wave over the length scale of several lattice constants, which might not be valid, depending on the initial and final photo-emission states. Additionally at the polycrystalline and possibly reconstructed and relaxed surfaces [53], we are dealing with in our experiments, the symmetry is reduced and a priori not known. We will therefore neglect any Bragg reflections in the course of this thesis.

4.2 Inelastic Scattering

Inelastic scattering in solids is modelled as the interaction of an electron with a medium characterized by the dielectric constant $\epsilon(\vec{r}, \omega)$. However, before looking into the details inelastic scattering calculations, we will consider the basics kinematics.

4.2.1 Kinematics of inelastic scattering

The kinematics are governed by the universal law of energy and momentum conservation:

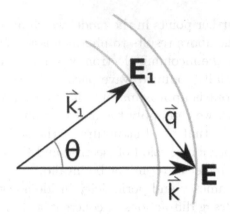

Figure 4.4: The kinematics of inelastic scattering.

$$E = E_1 + Q \tag{4.2}$$

$$\vec{k} = \vec{k_1} + \vec{q} \tag{4.3}$$

Where E ($\vec{k}$) and E_1 ($\vec{k_1}$) is the initial and final energy (wavevector) of the electron. Q is the energy loss and $\vec{q}$ the wavevector transfer. Rearranging $\vec{q} = \vec{k} - \vec{k_1}$, and squaring the expression leads to $q^2 = k^2 + k_1^2 - 2kk_1 cos\Theta$, where Θ is the scattering angle (see 4.4). To further simplify the expression, we have to make an approximation about the dispersion relation. Assuming a free-electron like dispersion $E = \frac{\hbar^2}{2m}k^2$ and using $E = E_0 - Q$, we arrive at the following relation:

$$q = \sqrt{\frac{2m}{\hbar^2}(2E - Q - 2\sqrt{E}\sqrt{E - Q}cos\Theta)} \tag{4.4}$$

That means that for a given energy loss Q there's a lower(-) and upper(+) bound for q:

$$q_{\pm} = \sqrt{\frac{2m}{\hbar^2}(\sqrt{E} \pm 2\sqrt{E - Q})} \tag{4.5}$$

By taking the total differential of Eq. 4.4 at fixed energy loss Q, we arrive at an expression which we will use later:

$$dq = \frac{2m}{\hbar^2} \frac{\sqrt{E}\sqrt{E-Q}sin\Theta}{q} d\Theta \qquad (4.6)$$

4.2.2 Theory

To arrive at the expression for the inelastic scattering cross section within the semiclassical framework some derivation is necessary, which is usually not found in the literature. The derivation we present here, has been adapted from the derivation of the energy loss of electrons in the presence of a surface [54]. We start by considering an electron with velocity $\vec{v}$ moving through a homogeneous medium characterized by a dielectric constant ϵ. We will neglect the magnetic field $\vec{B}$. This is equal to setting $\vec{A} = 0$, which is an approximation valid for non-relativistic particles. From Gauss-Law in frequency-momentum-space we obtain:

$$i\vec{q}\vec{D}(\vec{q},\omega) = \rho(\vec{q},\omega). \qquad (4.7)$$

Together with the definition of the potential in frequency-momentum-space $\vec{E}(\vec{q},\omega) = -i\vec{q}\phi(\vec{q},\omega)$ and the constitutive relation $\vec{D}(\vec{q},\omega) = \epsilon(\vec{q},\omega)\vec{E}(\vec{q},\omega)$, we can derive a result for $\phi(\vec{q},\omega)$. This can then be used to obtain $\vec{E}(\vec{q},\omega)$:

$$\vec{E}(\vec{q},\omega) = -\frac{i\vec{q}}{q^2\epsilon_0\epsilon(\vec{q},\omega)}\rho(\vec{q},\omega). \qquad (4.8)$$

The charge distribution $\rho(\vec{r},t)$ is given by $e\delta(\vec{r}-\vec{v}t)$, consequently $\rho(\vec{q},\omega)$ reads $2\pi e\delta(\vec{q}\vec{v}+\omega)$. Using $E_{ind} = E_{bulk} - E_{vac}$, we get an expression for the induced electric field:

$$\vec{E}_{ind}(\vec{q},\omega) = -2\pi\frac{i\vec{q}}{q^2\epsilon_0}e\delta(\vec{q}\vec{v}+\omega)\left[\frac{1}{\epsilon(\vec{q},\omega)-1}\right]. \qquad (4.9)$$

The easiest way to obtain a relation with the formalism of scattering cross sections is via the stopping power, the energy loss per unit path length in the medium. It reads as:

$$\frac{dW}{ds} = \frac{dt}{ds}\frac{dW}{dt} = \frac{1}{v}\vec{v}\vec{F}_{ind}\bigg|_{\vec{r}=\vec{v}t} = \frac{1}{v}e\vec{v}\vec{E}_{ind}\bigg|_{\vec{r}=\vec{v}t} = \tag{4.10}$$

$$-\frac{1}{v}\frac{\vec{v}e^2 2\pi}{\epsilon_0}\int_{R^3}\mathrm{d}q^3\int_{-\infty}^{\infty}\mathrm{d}\omega\frac{i\vec{q}}{q^2}\delta(\vec{q}\vec{v}+\omega)\left[\frac{1}{\epsilon(\vec{q},\omega)}-1\right]e^{+i\omega t}e^{+i\vec{q}\vec{r}}\bigg|_{\vec{r}=\vec{v}t}$$

Now there are two small tricks. First of all we move $\vec{v}$ into to integral and perform the scalar product with $\vec{q}$. Using the delta function this can be replaced with ω. Secondly, we notice that the expression is evaluated at $\vec{r} = \vec{v}t$. Again with the help of the delta function, the complex exponentials from the Fourier transform reduce to unity, resulting in:

$$\frac{dW}{ds} = \frac{1}{v}\frac{2\pi e^2}{\epsilon_0}\int_0^{\infty}\frac{\mathrm{d}q\,q^2}{(2\pi)^3}\int_{-1}^{1}\mathrm{d}\cdot\cos\Theta\int_0^{2\pi}\mathrm{d}\phi \tag{4.11}$$

$$\cdot\int_{-\infty}^{\infty}\frac{\mathrm{d}\omega}{2\pi}\left[\frac{1}{\epsilon(\vec{q},\omega)}-1\right]\frac{-i\omega}{q^2}\delta(qv\cos\Theta+\omega)$$

Assuming that the dielectric constant is isotropic, meaning $\epsilon(\vec{q},\omega) = \epsilon(q,\omega)$, we can perform the integral over ϕ to give a factor 2π. The integral over $\cos\Theta$ can be used to collapse the delta function, which gives an factor $\frac{1}{qv}$ and reduces the integral boundaries:

$$\frac{dW}{ds} = \frac{1}{v^2}\frac{e^2}{(2\pi)^2\epsilon_0}\int_0^{\infty}\mathrm{d}q\frac{1}{q}\int_{-qv}^{+qv}\mathrm{d}\omega(-i\omega)\left[\frac{1}{\epsilon(\vec{q},\omega)}-1\right]. \tag{4.12}$$

Using the identity $\epsilon(\vec{q},\omega) = \epsilon^*(-\vec{q},-\omega)$, we obtain the final expression:

$$\frac{dW}{ds} = \frac{1}{v^2}\frac{e^2}{2(\pi)^2\epsilon_0}\int_0^{\infty}\mathrm{d}q\frac{1}{q}\int_0^{+qv}\mathrm{d}\omega\,\omega\mathrm{Im}\left(\frac{-1}{\epsilon(\vec{q},\omega)}\right). \tag{4.13}$$

So far, we have a expression for the stopping power based on Maxwell's equations. Now we want to have another one based on differential cross-sections $\frac{d\sigma}{d\omega d\vec{q}}$. Using the target number density $\mathcal{N}$ the energy loss per unit path length can be expressed as

$$\frac{dW}{ds} = \int_0^E d\omega \int_{q_-}^{q_+} d\vec{q}\,\omega\mathcal{N}\frac{d\sigma}{d\omega d\vec{q}}. \tag{4.14}$$

The product of differential cross section and target number density in the kernel can be interpreted as probability per unit path length to loose an energy ω and momentum $\vec{q}$. Note that the integral is over all possible $\vec{q}$ and not a volume integral. Using the symmetry of our model, we can perform the integral over the angles:

$$\frac{dW}{ds} = \int_0^E d\omega \int_{q_-}^{q+} dq \int_{-1}^{+1} d\cos\Theta \int_0^{2\pi} d\phi\,\omega\mathcal{N}\frac{d\sigma}{d\omega d\vec{q}}$$

$$= 4\pi \int_0^E d\omega \int_{q_-}^{q+} dq\,\omega\mathcal{N}\frac{d\sigma}{d\omega dq}. \tag{4.15}$$

The quasi-classical approximation consists now in identifying the Fourier components ω and q of Eq. 4.13 with the energy loss and momentum transfer in Eq. 4.15. Comparing these two equations we obtain:

$$\mathcal{N}\frac{d\sigma}{d\omega dq} = \frac{e^2}{(2\pi)^2\epsilon_0 v^2} \cdot \frac{1}{q}\text{Im}\left(\frac{-1}{\epsilon(q,w)}\right). \tag{4.16}$$

Note that the integral limits in Eq. 4.15 come from kinetic considerations whereas in Eq. 4.13, we didn't impose such restrictions yet. Using the inelastic mean free path $\lambda = 1/\mathcal{N}\sigma$ and the electron energy E, the above expression can be written as

$$\frac{d\lambda^{-1}}{dQdq} = \frac{1}{a_B\pi E}\frac{1}{q}\text{Im}\left(\frac{-1}{\epsilon(q,Q)}\right). \tag{4.17}$$

Eq. 4.17 is the central equation in this chapter. It divides the scattering in a projectile dependent term and a material dependent term, the energy loss function (ELF) $\text{Im}(-1/\epsilon(q,Q))$. From this we can derive an equation for the so called differential inverse inelastic mean free path DIIMFP, which corresponds to the energy loss probability:

$$\frac{\mathrm{d}\lambda^{-1}}{\mathrm{d}Q} = \frac{1}{a_B \pi E} \int_{q_-}^{q_+} \frac{\mathrm{d}q}{q} \mathrm{Im}\left(\frac{-1}{\epsilon(q,Q)}\right) \tag{4.18}$$

and further the differential cross section using Eq. 4.6:

$$\frac{\mathrm{d}\lambda^{-1}}{\mathrm{d}\Omega} = \int_0^E \mathrm{d}Q \frac{\mathrm{d}\lambda^{-1}}{\mathrm{d}q\mathrm{d}Q} \frac{\mathrm{d}q}{4\pi \mathrm{d}cos\Theta} \tag{4.19}$$

4.2.3 Extension algorithms

In the presented model, the material dependent term, which determines all the scattering characteristics of a given solid is determined by $\epsilon(q,Q)$. This function is experimentally hard to access except at $q = 0$, where it can be calculated from optical data. Fig. 4.5 shows the energy loss function for $q = 0$ for gold. With knowledge of the electronic structure of the material it is possible to assign specific loss channels to certain features in the ELF. The extension of the ELF to non-zero q is difficult and has to rely on semi-empirical algorithms. (An overview is given in [55]). First principle calculations still have difficulties in reproducing the experimental data for $\epsilon(0,Q)$(see e.g. [56]). For a few model systems $\epsilon(q,Q)$ is approximately known analytically, for example the homogeneous electron gas. However these models are often not sufficient to describe the complex structure in real solids. The most successful strategy for real materials is the Optical Data Model. It parametrizes the ELF at $q = 0$ and varies some of the parameters with q according to semi-empirical models. The two most commonly used algorithms are the extended Drude model and the Drude-Lindhard model.

In the extended Drude model the dielectric function $\epsilon = \epsilon_1 + i \cdot \epsilon_2$ is expressed as a sum of Drude-terms (see Eq. 2.11):

$$\epsilon_1 = \epsilon_b - \sum_j \frac{f_j(\omega^2 - \omega_j^2)}{(\omega^2 - \omega_j^2)^2 + \omega^2\gamma_j^2} \tag{4.20}$$

$$\text{and } \epsilon_2 = \sum_j \frac{f_j\gamma_j\omega}{(\omega^2 - \omega_j^2)^2 + \omega^2\gamma_j^2}. \tag{4.21}$$

The imaginary part ϵ_2 is fitted to the experimental ELF from which ϵ_1 can be calculated. In order to extend this model to non-zero q's, ω_j is varied as:

$$\omega_j(q)^2 = \omega_j(0)^2 + \frac{2}{3}E_f\frac{\hbar^2}{2m_c}q^2 + (\frac{\hbar^2}{2m_e}q^2)^2. \tag{4.22}$$

where E_f is the Fermi energy, linked to the plasmon energy $E_p = \hbar\omega_0$ by $E_f = \frac{\hbar^2}{2m_e}(3 \cdot \pi^2\frac{\epsilon_0 m_e}{\hbar^2 e})^{\frac{2}{3}} \cdot E_p^{\frac{4}{3}}$. Often the simplified asymptotic version for high q is used:

$$\omega_j(q) = \omega_j(0) + \frac{\hbar^2}{2m_e}q^2. \tag{4.23}$$

Only sometimes also the damping constant is dispersed $\gamma = \gamma(q)$ [55]. The approach in the Drude-Lindhard model is different. Here the optical ELF is directly expressed as a sum of loss functions derived from Drude model:

$$\text{Im}\left(\frac{-1}{\epsilon(q,Q)}\right) = \sum_j f_j\text{Im}\left(\frac{-1}{\epsilon_j(\omega, q, \omega_j, \gamma_j)}\right) \tag{4.24}$$

$$\frac{-1}{\epsilon_j} = \frac{\gamma\omega_i(0)^2\omega}{(\omega^2 - \omega_j(q)^2)^2 + \gamma^2\omega^2}, \tag{4.25}$$

where $\omega_j(q)$ is extended to non-zero q as in the model above. One possibility to avoid the semi-empirical extensions of $\omega(q)$ is to use the Mermin-formalism [52], however paying the price of considerably increased complexity. In analogy to the Drude-Lindhard model the loss function is expressed as:

$$\text{Im}\left(\frac{-1}{\epsilon(\omega,q)}\right) = \sum_j f_j \text{Im}\left(\frac{-1}{\epsilon_m(\omega,q)}\right), \qquad (4.26)$$

where ϵ_m is the Mermin-function [57], which is the self-consistent description of the dielectric function of a homogeneous electron gas with a finite plasmon width in the Random-Phase-Approximation (RPA). It is based on the Lindhard dielectric function ϵ_L:

$$\epsilon_m(\omega,q;\gamma,E_f) =$$
$$1 + \frac{(1 + i \cdot \gamma/\omega)(\epsilon_L(\omega + i\gamma, q; E_f) - 1)}{1 + i(\gamma/\omega)(\epsilon_L(\omega + i\gamma, q; E_f) - 1)/(\epsilon_L(0, q; E_f) - 1)} \qquad (4.27)$$

The Lindhard dielectric function itself is given by [58]:

$$\epsilon_L(\omega,q;E_f) =$$
$$1 + \frac{k'_{tf}}{4q'^3}\left(2q' + (1 - x_1^2)\log\frac{x_1 + 1}{x_1 - 1} + (1 - x_2^2)\log\frac{x_2 + 1}{x_2 - 1}\right), \qquad (4.28)$$

where $x_1 = \frac{1}{2}(q' - \omega'/k')$ and $x_2 = \frac{1}{2}(q' + \omega'/q')$ and $k_{tf} = \sqrt{\frac{me^2k_f}{\epsilon_0 pi^2\hbar^2}}$ is the Thomas-Fermi wave-vector. The prime denotes that the quantities are expressed in Fermi units. Although it usually separated into its' real and imaginary parts for real valued arguments, it is advantageous for the complex arguments encountered in the definition of the Mermin function, to stay with Eq. 4.28 and use the complex logarithm.

In the limit $q \to 0$ the function $\text{Im}(1/\epsilon_m)$ reduces to the Drude-Lindhard expression (Eq. 4.24) [59]. Thus the same parameters can be used in parametrizing the ELF, only that now dispersion and broadening are naturally accounted for.

Although the formalism is recognized as a significant improvement over the extended Drude and Drude-Lindhard formalism in the low energy regime (e.g. [54,55]), it is not that often used. The reason besides the increased complexity is probably that as $q \to 0$ the evaluation of Eq. 4.27 becomes numerically unstable (because of the q^3-term in the denominator). We circumvent this problem by using the Drude-Linhard expression below an empirically determined q_{cut}.

It must be noted that ϵ_m is derived from the homogeneous electron gas. Thus it does not realistically describe all loss channels in a real solid, e.g. interband transitions. Often the occurrence of a band gaps, e.g. in semi-conductors is ignored, which is justified as long as the energy of the electron is much bigger the band gap([55]). One general possibility is to use Drude-Lindhard terms with a specific dispersion and broadening for such transitions and Mermin-terms for losses connected to plasmons.

One disadvantage of the extended Drude and Drude-Lindhard approach is that the ELF has to be parametrized through a complicated fitting procedure. There are other approaches which use integral equations, which however are not able to incorporate dispersion and broadening of the plasmon peak on the same level as the Mermin-formalism. Here we also use one formulation due to Ding and Shimizu [60], and refer to [61] for an overview of other possibilities.

4.2.4 Energy Loss Function

The first step in computing the inelastic scattering properties in solids is to fit the model to the optical loss function. There are a few databases available (see e.g. [56,62,63]). As can be seen from Fig. 4.5 a) the loss function computed from those differ quite considerably, although common features can be identified. We only mention, that there exist sum-rules to check the consistency of the input data (see e.g. [64]). We here choose to employ the data from [62], which is shown together with the corresponding Drude-Lindhard fit in Fig. 4.5 b). The fit was performed in the region from 0 to 190 eV. In the range from 10 to 100 eV it is quite accurate, below 4 eV and above 150 eV there are some small deviations.

Fig. 4.6 shows the extension of the loss function to non-zero momentum transfer computed with the Mermin-formalism. At $q = 0$

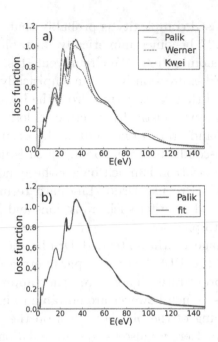

Figure 4.5: the loss function of gold: a) computed from different data
sources (green/thick dashes from [56], blue/solid line from
[62] and red/fine dashes from [63]) b) data and fit to [62] via
the drude-lindhard-formalism 4.24

we recognize the optical loss function. For higher energy losses and
momentum transfers a ridge can be observed, the so-called Bethe-
ridge, which asymptotically approaches a dispersion of q^2. Physically
that means that for high energy and momentum transfers the inelastic
scattering can be treated as being caused by interaction with free
electrons [65].

To check the influence of the inelastic scattering properties on the
employed model, we performed calculations of the inelastic mean free
path and the differential cross section with the Drude-Lindhard model,
the Mermin-model and the fitless procedure of Ding and Shimizu [60].
Gaussian quadrature was used for the numerical integrations. The

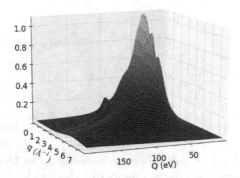

Figure 4.6: The extension of the surface loss function to non-zero momentum transfer computed through the Mermin-formalism.

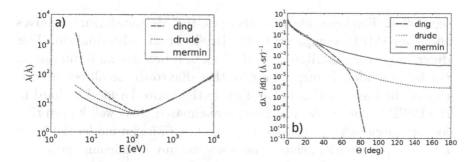

Figure 4.7: Comparison of the different extension algorithms: a) imfp b) differential cross section for incident electron energy of 100eV. At low energies differences occur while the models converge for higher energies

resulting inelastic mean free paths are shown in 4.7 a). The general behaviour is the same for all three models. Coming from high electron energies the IMFP decreases, reaching a minimum at around 100eV and then rising again. This behaviour is quite universal to all metals [66]. It can be seen that for high electron energies, the models

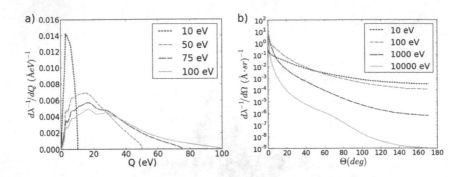

Figure 4.8: a) The differential inverse inelastic mean free path for gold
for different kinetic energies of the incident electron computed through the Mermin-formalism. b) The inelastic differential cross section for gold for different energies of the
incident electron.

agree well. For lower electron energies, the Mermin-formalism gives
the lower IMFPs compared to the Drude-Lindhard-formalism. This
difference can be attributed to the lack of a broadening formalism in
the latter model. It may be noted, that due to the small overestima-
tion of the loss-function below 4 eV in the above Drude-Lindhard fit
the IMFP below 10 eV might be overestimated. It is well known that
for low-energy electron scattering the Ding-Shimizu-model (DSM) is
a poor approximation, since it does not use any broadening formalism
and only a q^2-dispersion [60].

We also computed the differential cross section (DCS) for all three
models for an electron kinetic energy of 100 eV. As shown in Fig. 4.7
b), inelastic scattering is dominated by low-angle scattering. This
is intrinsic in this description of inelastic scattering, owing to the
$1/q$-term in 4.17. Due to the restrictions in the DSM, scattering with
an angle greater than $\pi/2$ is not allowed.

Furthermore, we calculated the energy loss probability and the
differential cross section within the Mermin model. As can be seen in
Fig. 4.8 a) the peaks from the optical loss function in Fig. 4.5 can
be recognized in the DIIMFP. The loss probability however is broad

and extends from low energies to the kinetic energy of the electron. For other materials, especially free-electron-like metals such as Al and Mg, the loss function is dominated by a single plasmon peak. This single plasmon peak can be observed even with a broadband XUV excitation and has even been measured with attosecond streaking [41].

The variation of the DCS is shown in Fig. 4.8 b). The general small-angle scattering characteristic already mentioned above, increases with energy.

There exist a number of databases, most prominently from NIST [67]. They usually only provide IMFPs, and don't give DIIMFPs and DCSs which are necessary for Monte Carlo simulations. This is the reason why we took the effort of implementing the above algorithms.

It must be noted that there exist a number of analytic expressions for the IMFPs obtained through a fit to numerical calculations. As we have seen above, a simple extension of this expression to low energies below 40 eV, where the IMFPs calculated from different models can differ by factors of a few 100%, should only be taken with care.

However at such low energies, expression 4.17 and the possibly also the model for the loss-function should be modified anyhow, to include exchange and correlation effects. ([68], [65]). This is beyond the scope of this thesis.

4.3 Surface Scattering

As the electron approaches the surface it interacts with the surface plasmon modes. This interaction is dependent on the distance to the surface and leads to a change in the energy loss probability compared to the bulk. An extensive literature exists on this problem for plane surfaces (see [54] for an excellent introduction and overview). For nanostructures the spatial dependence of the surface loss has been applied to measure the spatial distribution of different nanoplasmonics modes using electron energy loss spectroscopy EELS (e.g. [69]). For the example of a sphere the connection between electron energy loss

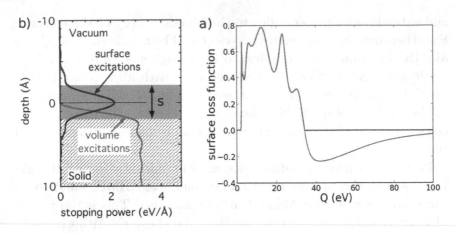

Figure 4.9: a) Illustration of surface and bulk surface losses. b) The surface loss function for $q = 0$ of gold computed from [62].

and plasmon modes can be understood from the derivation of Eq. 2.32 and Eq. 4.17. In the framework of Mie theory the fields in any region excluding the electron can be described through the vector harmonics. Solving for the boundary conditions gives the induced fields acting on the electron in terms of the different modes [70]. The difference to light scattering is, that the electron will generally also couple to non-dipole, so called dark, modes.

In the nanoplasmonic streaking experiments considered in this thesis, we will encounter different shapes. For a general description we will use an approximation where we consider the surface as being (locally) flat and investigate the scattering characteristics. Since the general description [54] depends on five parameters, energy E, depth d, angle to surface α and scattering angles θ and ϕ, and leads to expressions which are very hard to evaluate, we will use a simplified description, given by [63]. In this model it is assumed, that the region where electrons couple to the surface modes is usually small (see Fig. 4.9 a)). Assuming scattering to be confined to the plane of incidence and straight line propagation through this region (SLA), effectively integrating over the surface loss term, this expression can be obtained:

$$P_s(\alpha, E, Q, \vec{q}) = \frac{|q_s|}{\pi^2 v \cos\alpha q^4} \mathrm{Im}\left(\frac{(\epsilon - 1)^2}{\epsilon(\epsilon + 1)}\right), \qquad (4.29)$$

where P_s is the surface contribution of electrons scattering and losing an amount of Q in energy, α the crossing angle and q_s the momentum transfer parallel to the surface. The same relations between E, Q and θ and q as for inelastic scattering in the bulk are valid.

The material dependent term in this equation is given by the surface loss function:

$$\mathrm{Im}\left(\frac{(\epsilon - 1)^2}{\epsilon(\epsilon + 1)}\right). \qquad (4.30)$$

As can be seen in Fig. 4.9 b), the SLF is also negative in certain regions, where the bulk loss function is maximal.

Further assuming that the scattering is confined to the plane of incidence, q_s can be decomposed as, depending of whether the electron is scattered from or to the surface [63]:

$$q_s = q_\perp \cos\alpha \pm q_\| \sin\alpha = \left(q^2 - \left(\frac{Q}{v} + \frac{q^2}{2v}\right)^2\right)^{\frac{1}{2}} \cos\alpha \pm \left(\frac{Q}{v} + \frac{q^2}{2v}\right)\sin\alpha. \qquad (4.31)$$

With this, the so called surface excitation parameter (SEP), the mean number of surface excitations per crossing, can readily be calculated:

$$P_{s\pm}(\alpha, E) = \frac{1}{\pi v^2 (\cos\alpha)} \int_0^E dQ \int_{q_-}^{q_+} dq \frac{|q_s'|}{q^3} \mathrm{Im}\left(\frac{(\epsilon - 1)^2}{\epsilon(\epsilon + 1)}\right). \qquad (4.32)$$

For the total SEP, both plus and minus-contributions have to be added:

$$P_s = P_{s-} + P_{s+}. \qquad (4.33)$$

Fig. 4.10 shows the calculated dependence of the SEP on the incidence angle and the energy. For low angles the SEP stays relatively constant and then diverges for angles approaching $\pi/2$. From the assumption that the SEP is proportional to the time the electron needs to cross the boundary, a simple formula for the SEP can be derived [71]:

$$P_s \propto \frac{s}{v \cos\alpha} = \frac{a_s}{\sqrt{E}\cos\alpha}, \qquad (4.34)$$

where a_s is a material specific parameter. As can be seen in 4.10, this approximation describes the variation of the SEP very well. For a_s in our region of interest of 30 eV to 140 eV, we obtain through fitting parameter values of about 2.5 to 2.8.

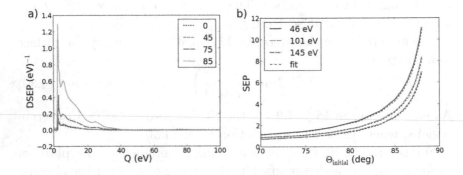

Figure 4.10: a) The differential surface excitation parameter (DSEP) for gold for different incidence angles. b) The surface excitation parameter (SEP) for gold. Dashed lines show the fit through Eq. 4.34.

The presented model is also capable of predicting energy losses via the so called differential surface excitation parameter DSEP. This is shown in Fig. 4.10 a) for an electron energy of 100 eV. While the angle dependence of the total area is approximately given by Eq. 4.34, the relative shape does practically not change. Due to the negativity of the the SLF, also the DSEP is slightly negative in some regions. This is no problem if the above expression is used for deconvolution of electron spectra. For Monte-Carlo simulations this negativity hinders the interpretation of the SEP and DSEP as probabilities. In our approach we just set the SLF to zero where it is actually negative. The expressions calculated by this procedure, turn out to practically not differ from the otherwise computed quantities in the case of gold. We note that for other materials, like Al, this approach might not be feasible. Moreover, the effect of this approximation on the angular distribution is not clear.

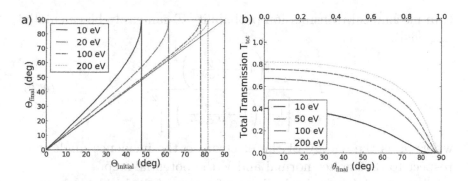

Figure 4.11: a) The relation between incidence and final angle in the transmission through a potential step. b) The effective total transmission T through the potential step dependening on final angle.

For the description of photoemission, electrons might actually be born inside the surface scattering region. For those electrons this model will overestimate the total scattering probabilities. Furthermore the assumptions of a straight-line propagation and scattering in the plane of incidence might not be valid. A consideration of the first mentioned idealization will lead to a reduction of the SEP for large incidence angles [71].

Being aware of the shortcomings of this model, we will still use it, for its qualitatively complete description of the inelastic surface scattering process.

4.4 Transmission

A point which is usually omitted in the context of inelastic surface scattering of electrons is the potential the electron has to overcome at the surface. Using the simplified model of a potential step, the quan-

tum mechanical transmission probability T can easily be calculated
(e.g. [64]):

$$T = \frac{4\left(1 - \frac{V}{E\cos^2\alpha}\right)^{\frac{1}{2}}}{\left[1 + \left(1 - \frac{V}{E\cos^2\alpha}\right)^{\frac{1}{2}}\right]^2}, \tag{4.35}$$

where E is the energy of the electron, α the incidence angle with
respect to the surface normal and V the potential step.

The potential step also leads to refraction of the incident electrons.
The relation between initial and final angle α and α' is given by:

$$\sin\alpha' = \left(\frac{V}{E - V}\right)^{\frac{1}{2}} \sin\alpha \tag{4.36}$$

and shown in Fig. 4.11.

In a more or less ad-hoc approach we can try to combine both
surface scattering effects. Assuming that the inelastic scattering
region can be divided into two equal regions above and below the
surface and considering the Poissonian distribution of the surface
excitation process [63], we can write for the probability of an electron
to leave the surface unscattered in dependence on the final angle α'
and initial energy E:

$$T_{tot}(\alpha', E) = \exp\left(-P_s(\alpha', E - V)/2\right)T(\alpha, E)\exp\left(-P_s(\alpha, E)/2\right) \tag{4.37}$$

where we used Eq. 4.36 for T and the analytic approximation 4.34 for
P_s. The result is plotted in Fig. 4.11 for different energies without
(dashed) and with (solid) inelastic surface scattering. It can be seen
that the probability for unscattered transmission through the surface
decays abruptly for large final angles. This will affect the weight of
different photoelectron contributions coming from different regions of
the nanoobject.

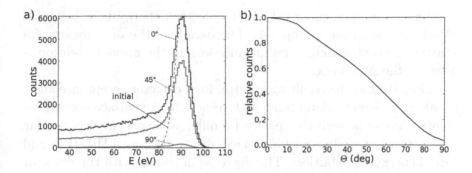

Figure 4.12: a) Electron kinetic energy spectra for different detector orientations and a detector acceptance of 40°. While the shape doesn't change much, the countrates vary considerably. b) Angular dependence of the total electron counts for detector acceptance of 40°

4.5 Simulation for a plane surface

To investigate the effect of the different scattering contributions, we performed a Monte-Carlo-simulation of electron transport at a gold surface. Therefore we initialized electrons equally with different distances to the surface, an isotropic velocity distribution and a Gaussian distribution of initial kinetic energies around 95 eV with a FWHM of 8 eV. We assume free electron dispersion.

The transport in bulk is described by the total electron mean free path:

$$\frac{1}{\lambda_{tot}} = \frac{1}{\lambda_{inel}} + \frac{1}{\lambda_{el}}. \tag{4.38}$$

In this Monte-Carlo formulation, the distance ds to the next scattering event is given by:

$$ds = -\lambda_{tot}\log(1 - P), \tag{4.39}$$

where P is a uniform random number between 0 and 1. If the location of the next scattering event lies outside the surface, the electron is transmitted through the surface. The final angle to the surface, as well as kinetic energy, time for reaching the surface and number of scattering events is recorded. If the electron energy falls below 35 eV the propagation is stopped. The details of the description of a scattering event through the formulations in the previous section is given in the Appendix.

In Fig. 4.12 a) the resulting spectra, for a detector acceptance angle of 40° for different orientations with respect to the surface normal are shown. As can be seen the spectra for different angles are very similar. No distinct loss peak can be identified, due to the broad DIIMFP- and initial energy distribution. The decrease of counts with the detector angle can be qualitatively understood by considering bulk scattering alone. The inelastic mean free path λ_{inel} limits the average distance an electron can travel before losing too much energy. This leads to a reduction of the active volume $V_{emission}$ from which electrons can reach the surface depending on the angle to the surface α, which is given by:

$$V_{emission} \propto \lambda_{inel} \cdot \cos\alpha. \tag{4.40}$$

The monotonic decrease predicted by this simplified model is indeed encountered Fig. 4.12 b). This effect has so far not been considered in nanoplasmonic streaking simulations. Of course, in experiments, also polarization and initial state effects as well as the existence of surface states, might alter this result. Also the role of surface roughness is not clear.

The number of inelastic, elastic and surface scattering events for particles reaching the surface are shown in Fig. 4.13 a). The ratio of the number of inelastic and elastic scattering events is in good agreement with the ratio of the inverse mean free paths.

Another very important result is shown in Fig. 4.13 b). 63% of the electrons in the whole energy range from 35 eV reach the surface within 73 as. That means, that in this model scattering does indeed not lead to huge distortions in the timing and that the assumption of immediate emission in nananoplasmonic streaking ([25–28]) can be considered valid.

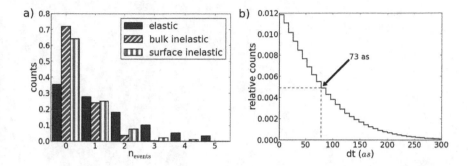

Figure 4.13: a) Number of elastic, inelastic and surface scattering events
for electrons reaching the surface with a kinetic energy
above 35eV. b) The delay time of electrons at the surface
after initialization for electrons with a kinetic energy above
35eV.

5 Attosecond streaking from metal nanotips

In this section experimental results of attosecond streaking experiments from a nanoplasmonic system, namely a gold nanotip, are presented. First, we will discuss some characteristics of nanoplasmonic streaking which arise due to the inhomogeneity of the electromagnetic fields. Then we will look into the theoretical modelling and expectations of our experiment. This subsequently allows us to better interpret and discuss the experimental results. Finally we will theoretically consider a slight modification of the experimental setup, with which we expect to easily obtain a clear signature of nanoplasmonic streaking.

5.1 General characteristics of nanoplasmonic streaking

Unlike in streaking from a gas (see chapter 2) the field around a metal nanoparticle is inhomogeneous due to localization and field enhancement. This is schematically shown in Fig. 5.1. The momentum shift of the electrons due to the streaking field is now given by:

$$\Delta \vec{p} = -e_0 \int_{t_0}^{\infty} dt \vec{E}(\vec{r}, t). \tag{5.1}$$

where the electric field has now become position dependent and the simple relation given in Eq. 2.37 is not valid anymore. This complicates the reconstruction of the near-field at the surface of the nanoparticle. Depending on the ratio of the field decay length κ and

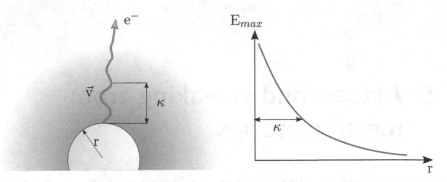

Figure 5.1: Illustration of the near-field decay around plasmonic
            nanoparticles. The decay length κ can usually be approx-
            imated by the size of the geometrical features r of the parti-
            cle. The regime of streaking will depend on the ratio of the
            initial electron velocity and the decay length of the field.

the initial velocity of the electron three regimes can be distinguished.
In the first extreme case, the decay length is so long that the electron
does not experience any position dependent variations before the pulse
has decayed in time. The electric field can therefore be considered
homogeneous, which is equivalent to the gas streaking case and is
therefore called ponderomotive regime. In the other extreme case, the
field decay length is so small, that the electron leaves the influence of
the near-field within a fraction of the pulse-duration. This is called
field-probing regime, the reason for this will be clear later. The third
regime, where neither of those assumptions is valid, is called the
intermediate regime. An intuitive model, which allows the distinction
of these regimes has recently been presented in [33, 45]. In this model,
the electric field is assumed to decay exponentially in space and time:

$$E(r,t) = E_0 e^{i\omega \cdot t} e^{-t/\tau_t} e^{-r/\kappa}. \tag{5.2}$$

where ω is the central frequency of the near-field oscillations, τ_t and
κ is the temporal and spatial decay constant, r is the distance to the
surface. The field is assumed to be normal to the surface.

For usual streaking amplitudes the change in the velocity v of the electron can be neglected when regarding the propagation in space [45]. This implies that for an electron propagating away from the surface with an initial velocity v, the spatially dependent part in the description of the electric field experienced by the electron may be rewritten as:

$$\frac{r}{\kappa} = \frac{v(t-t_0)}{\kappa} = \frac{(t-t_0)}{\tau_s}.$$ (5.3)

where $\tau_s = \frac{\kappa}{v}$. With this Eq. 5.1 can be integrated analytically:

$$\Delta p = -e_0 \int_{t_0}^{\infty} dt \, E(r(\vec{t}), t) = -e_0 \int_{t_0}^{\infty} dt E_0 e^{(i\cdot\omega - 1/\tau_t - 1/\tau_s)\cdot t} e^{+t_0/\tau_s} =$$

$$- e_0 E_0 \left[\frac{1}{i\omega - 1/\tau_t - 1/\tau s} e^{(i\cdot\omega - 1/\tau_t - 1/\tau_s)\cdot t} \right]_{t_0}^{\infty} e^{+t_0/\tau_s} =$$

$$E_0 \left[e_0 \frac{-i\cdot\omega - (1/\tau_t + 1/\tau_p)}{\omega^2 + (1/\tau_t + 1/\tau_s)^2} \right] e^{(i\cdot\omega - 1/\tau_t)\cdot t_0}.$$ (5.4)

The factor in the squared brackets defines the relation to the electric near-field at the surface (r=0) of the particle (compare Eq. 5.2). It can be factorized into a phase ϕ and amplitude A contribution:

$$\phi(\omega, \tau_t, \tau_s) = -\pi + arctan\left(\frac{\omega}{1/\tau_t + 1/\tau_s}\right).$$ (5.5)

$$A((\omega, \tau_t, \tau_s)) = \frac{1}{\sqrt{\omega^2 + (1/\tau_s + 1/\tau_t)^2}}.$$ (5.6)

Taking the limit $\kappa \to 0$, we see that the phase-shift to the electric field vanishes, which is why this regime is called field-probing regime.

In order to see the effect of propagation, on the measured momentum shifts, they are compared to the case of a completely homogeneous regime ($\kappa \to \infty$), depending on the initial energy of the electrons and the spatial decay constant. The result of such an analysis is plotted in Fig. 5.2 assuming a central wavelength of 720 nm and a temporal decay constant of 4.5 fs. Since τ_s also enters in the homogenous regime, it only becomes important when it gets below the period of

the field oscillations. The relative phase is shown in Fig. 5.2 a), where
the phase shift is expressed in units of time using ω. Only for a very
small spatial decay lengths below 10 nm and high electron energies
above a few keV, the phase shift approaches $\pi/2$ (fieldprobing regime),
whereas it stays below 50 as for κ >20nm and energies below 200
eV (ponderomotive regime). The shorter the eletron stays within the
influence of the near-field, the less it is accelerated during propagation.
This an be seen in Fig. 5.2 b). For the field-probing regime, the
relative amplitude approaches zero. In the ponderomotive regime by
contrast the relative amplitude is close to unity.

We refer to a more complete discussion of this model [45]. As
can be seen in the next section, we are experimentally dealing with
photoelectrons at around 90 eV and nanoobjects whose geometric
features are on the order of 50 nm. From Fig. 5.2 it an be seen that
it's thus save to assume that we are in the ponderomotive regime.

5.2 Theoretical Modelling

This section covers the modelling of the attosecond streaking experiments from a gold nanotip. The nanotips are provided by the
group of Peter Hommelhoff (FAU Erlangen). They are produced
from 100μm-thick polycrystalline gold wires by the so-called double-
lamellae drop-off technique, an electrochemical etching method. With
this technique tip-radii of a few 10 nm and a surface roughness below
0.8 nm have been realized [9, 72]. Fig. 5.3 a) shows a TEM-image of
one of the tips used in experiments. Based on this, the tip is modelled
as a semi-infinite cylinder with a semi-sphere (Fig. 5.3 b)) with a
radius of 100 nm as tip apex.

To get an expectation of the response of this system to a fs-IR-pulse
this geometry has been modelled numerically using a commercial
finite-difference time-domain (FDTD) solver [73]. The input pulse
is chosen to be a Gaussian with an intensity-FWHM of 4.5 fs and a
center wavelength of 720 nm. The CEP is chosen to be 0°. Due to
memory restriction on the work station used for the calculations, the

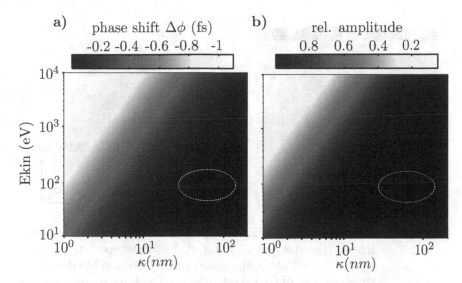

Figure 5.2: phase-shift (a) and relative amplitude (b) of streaking traces taking into account the spatial and temporal decay compared to the homogeneous case ($\kappa \to \infty$) but same temporal decay constant τ_t. Calculated for $\tau_t = 4.5 fs$ and $\lambda = 720 nm$ through Eqs.5.5 and 5.6. The dashed white circles mark the regime relevant for the experiments presented in this thesis.

simulation area is confined to a volume of 6 μm$\times$6 μm$\times$3 μm. As a consequence the diameter of the Gaussian input beam was limited to 4 μm in order to avoid numerical diffraction at the simulation region boundary. The dispersion for gold is taken from [74]. The mesh at the apex of the tip has a resolution of 2 nm.

Fig. 5.4 a) and Fig. 5.4 b) show the maximum of the electric field in the direction of the cylinder axis relative to the input pulse in the plane along and perpendicular to the laser propagation direction. At the tip apex the characteristic features of nanoplasmonics, field enhancement and field localization can be recognized. The field is localized down to a length scale on the order of the radius of the nanotip. For this relatively large apex the field enhancement is limited to a factor of about 2. The maximum field strength at the shank is

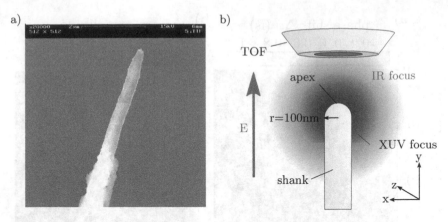

Figure 5.3: a): TEM-image of one of a gold nanotip used in the experiments [Courtesy of M. Förster]. b) Model geometry: The tip is oriented along the polarization direction towards the time-of-flight (TOF) spectrometer. The laser propagates into the paper plane.

reduced to a value of about 0.5 compared to the input pulse. This can be understood as arising due to the superposition of the incoming wave with a phase shifted reflected wave. Note again that for the whole geometry the decay length of the near-fields is on the order of 50-100 nm and the considered streaking processes with electron energies of around 100 eV lies deep in the ponderomotive regime, as discussed above.

Analyzing the temporal profile of the eletric fields a different positions one finds a temporal phase shift between shank and tip apex. Additionally it can be observed that the temporal duration of the fields at the nanotips are practically the same as those of the input pulse and it can therefore be concluded that the response of the system is ultrafast. This absence of a sharp resonance in the field enhancement from nanotips is often used to argue, that the field enhancement is not a plasmonic but rather a geometrical effect. However as discussed in chapter 2, the collective electron motion in nanoplasmonics is partly due to the nanosize of the object and not necessarily due to the material property alone. We will therefore keep the term nanoplasmonic in this work and refer to [75] for a discussion.

In order to assess the signature of the near fields on the streaking traces, we calculate so-called pseudo-streaking traces. For different delays between XUV- and IR-pulse, electrons are emitted at specified points from the surface with a fixed energy and velocity, and subsequently propagated in the electric field using a velocity verlet algorithm [76]. The propagation is stopped at a time when the electric field has substantially decayed and the final kinetic energy, as would be measured by a TOF, is recorded. Fig 5.4 c) shows the result of such a calculation. The emission positions of the electrons are indicated in Fig. 5.4 a) and b) by colored dots. The electrons emitted from the apex show the highest streaking amplitude, which rapidly decays moving away from the position of maximum field enhancement. The near-fields at the shank of the tip, are closely related to the Mie-solution of a cylinder, and approximately independent of the distance from the apex, which can also be seen in the pseudo streaking curves. For comparison we also calculate the streaking curve for the same settings but without any object in the simulation volume. This corresponds to a gas streaking measurement and is termed 'reference' from now on. It is shown as a dashed curve in the above plot. The amplitude ratios between reference and tip signal agree quite well with the ratios of the maximum field strength.

In addition to the different streaking amplitudes, there is also a also shift observable between the peak time of the different contributions, which will be called peak shift in the following. Fig 5.4 d) shows the pseudo-streaking curves on a smaller time scale. While the shank contribution advances the reference by around 250 as, the apex contribution is delayed by approximately the same amount. It is important to note that such a shift is a priori not constant over the pulse due to material dispersion in combination with the geometry. As discussed in chapter 2, dispersion in the frequency domain is equivalent to a delayed response in time. The actual peak shift will then depend on the pulse shape and might vary over the pulse. Consider the extreme case of a sharp resonance in the dielectric function covered by the

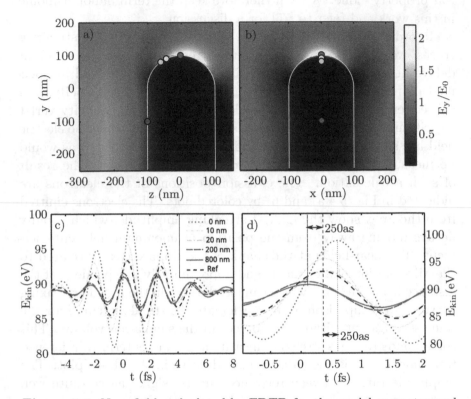

Figure 5.4: Near fields calculated by FDTD for the model geometry and
the signature in an attosecond streaking experiment: a) and
b) maximum of the electric field in direction of the cylinder
axis perpendicular and parallel to the laser propagation
direction for a 4.5 fs intensity-FWHM Gaussian input pulse
centred at 720 nm. The dots indicate the emission positions
of the electrons in the pseudo-streaking curves of shown in c)
and d). The legend indicates the distance from the apex in
y-direction. The last point lies outside the range of a) and b)
and is not shown.

bandwidth of the laser pulse but not at the central wavelength. This
will lead to a shift of the central frequency of the near-fields which

automatically implies a change of the peak shift between tip and reference. Regarding the response of shaped objects this has to be complemented with the geometrical considerations. Generally the same geometry will respond differently to different frequencies due to the change in the wavelength and the dielectric constant. The combined effect of the dielectric function and the geometry can conveniently be discussed in the framework of linear response theory. In the frequency-domain the response of the system at a point $\vec{r}$ to an input field $E_{input}(\omega)$ can be described via the complex valued response $H(\vec{r}, \omega)$

$$E(\vec{r}, \omega) = H(\vec{r}, \omega) \cdot E_{input}(\omega). \tag{5.7}$$

where $E(\vec{r}, \omega)$ is the total field due to E_{input}. If the linear response of the system and the spectral representation of the input field is known, the answer of the system in the time domain can be calculated via Fourier transformation. We note that in this framework the discussion of the resulting temporal evolution of the total field is completely analogous to the general formulation for a light pulse travelling through a linear dispersive medium [32]. Fig. 5.5 shows the amplitude and the phase) of the linear response at the front of the shank of the gold nanotip, 1 nm above the surface, for a polarization direction parallel to the cylinder axis. The model of an infinite cylinder was used and the response calculated by Mie theory [77]. Additionally the spectral intensity of a Gaussian laser pulse with an intensity-FWHM of 4.5 fs centred at 720 nm is shown. Both the amplitude and the phase are practically flat over the whole spectrum. That is the reason, why the peak shifts calculated in the time domain stay constant over the main and trailing part of the pulse and why, as observed above, the temporal response of the system is ultrafast.

We also note at this point that we employed a very simplified model and e.g. completely neglected the interaction of the photoelectron with the remaining hole and its quantum mechanical nature. Inclusion of this effect will lead to a temporal shift on the order of a few 10 as [33]. Furthermore a self-consistent description of attosecond streaking from metal surfaces under normal emission showed that considering the

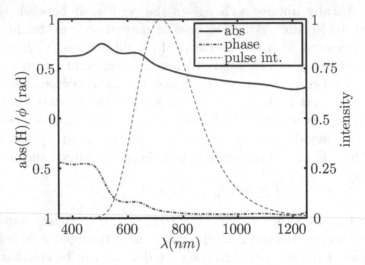

Figure 5.5: The linear response of the shank to the incident electric
field using the infinite cylinder model and calculated with
Mie theory [77] in terms of the amplitude (abs(H)) and the
phase (ϕ). The electric fields are parallel to the cylinder axis.
The response is flat over the whole spectrum. The relative
amplitude and the peak shift calculated with the phase of
the response at the central wavelength agree well with the
FDTD calculations performed in the time domain. The
spectral intensity of a 4.5 fs-pulse centred at 720 nm is also
shown

dielectric interaction of the electron with the solid in a quantum
mechanical framework leads to a shift of about 100 as [78]. Under
large angle emission this effect might even be larger, which is relevant
for our geometry. The calculated phase shifts have to be understood
as a guideline for the expectations on experimental results.

For the simulation of the experimental streaking traces, a full
3D-calculation is performed using the IR-fields from above. The
fields are linearly interpolated between grid points. The experimental
XUV-spetrum is taken into account and a Gaussian shape with an
intensity-FWHM of 220 as is assumed. The XUV-propagation and

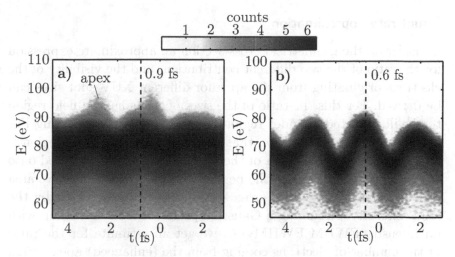

Figure 5.6: a): simulated attosecond streaking measurement from gold nanotips as described in the text. b) The simulated gas streaking reference trace

electron emission from the solid is based on the description in chapter 2. Due to limitations in the monitor size, emission is restricted to within 1.2 μm from the apex. The electron is then propagated under the influence of the IR-field using again a velocity-verlet algorithm. The dispersion is chosen to be free electron like. Electrons are born at the surface and scattering is neglected. The result of a full streaking calculation is shown in Fig. 5.6 a).

With the help of the pseudo-streaking curves one can identify the different contributions. The major part originates from the shank of the tip, while only few counts are emitted from the apex. The latter contribution can be identified by the shift of the peak position and higher amplitude. However it is only visible for positive energy shifts, as the biggest part of this contribution is hidden within the broad spectrum.

Count rate approximation

Considering the geometrical cross-section an approximate expression for the ratio of the two different contributions and the visibility of the electrons originating from the apex for different XUV spot sizes can be derived. For this the ratio of the sizes of the enhanced field region to the illuminated cylinder region has to be estimated. Taking Fig. 5.4 as a guideline we deduce, that the area of the enhanced region is about 10% of the total area of the semi-sphere, when projected onto the plane perpendicular to the beam direction. For the effective area of the shank, the cylinder cross section has to be weighted by the XUV intensity. Assuming a Gaussian profile of the XUV beam with an intensity-FWHM $FWHM_{XUV}$, we get an estimate for the ratio of the number of electrons coming from the (enhanced) apex region N_{apex} to the number of electrons originating from the shank N_{shank}:

$$\frac{N_{apex}}{N_{shank}} \approx \frac{0.1 \cdot \pi r^2}{2r \frac{\sqrt{\pi}}{2} \frac{FWHM_{XUV}}{2\sqrt{ln(2)}}} \approx 0.2 \frac{r}{FWHM_{XUV}}. \qquad (5.8)$$

where r is the radius of the tip. We note that this is only a rough estimate, e.g. we did not consider the decrease of amplitude between tip apex and beginning of the cylinder. If $N_{shank} >> N_{apex}$ we can approximate $N_{apex} = 0.2 \frac{r}{FWHM_{XUV}} \cdot N_{total}$. Whether all the electrons are visible will depend on the ratio of the streaking amplitude to the width of the photoelectron spectrum. For the simulation we get about $N_{apex} = 0.02 \cdot N_{total}$.

Different geometries

We also performed FDTD simulations for geometries which go beyond this idealized model for the tip. We checked cylinders which were tilted by an angle of $20°$ with respect to the polarization direction both in the plane of laser propagation and perpendicular to it. While the maximum field enhancement slightly varies, all phase relations and the amplitude relation between gas reference and shank turned out to be robust against this variations.

Realistic tips have a conical shape with opening angles on the order of 10° ([33] and see Fig. 5.3). The effect of the opening angle was studied with a conical tip with an opening angle of 15° and an apex radius of 50 nm. Again the relations appeared very robust, however only a small region within 2 μm from the apex could be simulated. For experimentally realistic XUV spot sizes of around 10 μm the radius of the conical tip changes by a few 100 nm within the XUV-focus. In order to estimate the effect of this on the streaking traces we used the observation that in the above FDTD-simulations of a conical tip, the results where close to the Mie-solution for infinite cylinders. Using the code from [77], the electric field in the direction of the cylinder axis at 1 nm from the surface was calculated again for a 4.5 fs input pulse centred at 720 nm. The dispersion for gold was taken from [62]. The electric field at 1 nm from the surface and the input pulse were integrated to give the 'local' vector potential. Relative amplitudes and peak shifts between the largest peak of each pseudo-potential were extracted. The results are shown in Fig. 5.7. The peak shifts are almost independent of the angle and slightly increase with the radius from 250 as at 50 nm to 450 as at 1500 nm. The relative amplitudes decrease with the angle but have a similiar shape. The maximum of the relative amplitude slightly reduces from 0.55 at 50 nm to 0.48 at 1500 nm. For both observables the change with the radius becomes smaller with bigger radii. From this analysis one can expect a rather homogeneous streaking trace, close to Fig. 5.6 a), even when using conical tips and relatively large XUV spot sizes.

5.3 Experiments

The first step and one of the central parts in the experiments is the optimization of the HHG-process and the generation of isolated attosecond pulses. This is done via streaking from a neon gas target which is placed in front of the TOF. Only then the nanotip is moved

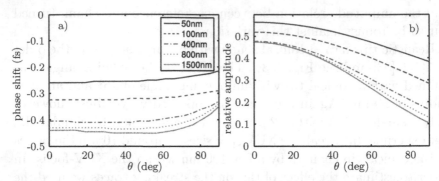

Figure 5.7: The relation of the vector potentials of the near-fields for
different radii to the incoming field. Calculations where
performed using a Mie-code for infinite gold cylinders [77]
for polarization parallel to the cylinder axis at 1 nm from
the surface. θ is the angle between radial vector and the
propagation direction of the laser. $\theta = 0°$ corresponds to the
front of the cylinder, $\theta = 90°$ to the side. a) shows the peak
shifts, b) shows the relative amplitude. Only minor shifts are
to be expected due to the conical shape of the tip and the
propagation of electrons away from the cylinder surface. Due
to the quasi-constant phase-shift with respect to the angle, a
homogeneous streaking trace can be expected.

into the laser focus. Fig. 5.8 a) shows an experimental spectrum of a
gold nanotip under the combined illumination by an IR- and an attosecond XUV-pulse. The spectrum is composed of two contributions.
First electrons which are produced by a linear photoemission process
through XUV-photons at a kinetic energy around 90eV (hereinafter
called XUV-electrons). Secondly electrons are produced due to nonlinear photoemission by the strong IR-field below an energy of about
15 eV, subsequently called IR-electrons (for an overview of nonlinear
photoemission from metal nanotips see [72]). Some electrons in this
region are also produced by scattering of XUV-electrons. As can
be checked by blocking the IR-beam, this contribution is, however,
negligible.

IR-electrons mainly originate form the tip apex, where the highest field enhancement is realized, due to the high nonlinearity in the photoemission process. As shown in Fig. 5.8 b) and c), by scanning the tip in the plane perpendicular to the laser propagation direction and observing the total IR-eletron yield, the tip apex can be positioned in the laser focus. The XUV-electron yield by contrast is due to its' linearity proportional to the total cross section of the illuminated area of the nanotip. The above procedure is used when first switching from the gas to the tip target and is a lengthy process. For later changes the tip is moved out of the focus only along the laser propagation direction, in order to be able to position it as reproducibly as possible with the open loop stages used for the positioning of the tip.

5.4 Analysis

After the optimization of the gas streaking traces the IR-intensity is reduced, which is necessary to avoid damaging of the nanotip. As a reference and in order to be able to extract laser intensity another gas scan is performed before the nanotip is moved into the focus. A streaking measurement from the tip is usually divided into several repeated scans covering the same delay range. Since one measurement can take several hours, this is done to obtain some useful data even in cases when e.g. the CEP drops out of the control loop. The single scans only contain relatively few counts and the shape of the streaking curve can usually only be estimated. Under stable conditions, after a sufficient number of scans the gas target is again moved in to record a second gas reference scan. This is used to judge the occurrence of any phase drifts or jumps and changes in intensity during the preceding measurement.

Consequently, the experimental data is divided into two different groups, depending on whether they contain one or two references. Double-reference measurements, which show a phase shift between the two reference scans, or single-reference measurements, where the coarse streaking trace between the different scans differs considerably,

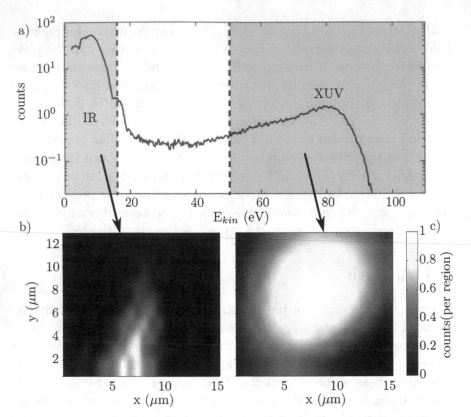

Figure 5.8: a): Experimental spectrum from a gold nanotip, the IR and
XUV parts discussed in the text are indicated. Note the logarithmic scale. b): The overall counts for the IR-part. c): The
overall counts for the XUV-part. The scales are extracted using the step size communicated by the manufacturer. Same
geometry as Fig. 5.3. [taken from [45]]

are discarded. With this our dataset reduces to a total of 9 measurements with a total of 26 peaks, 4 of those which comprise 12 peaks
are double-reference measurement.

We employ three different methods, in order to analyse the peak
positions and amplitudes of our streaking curves. The simplest one
is the so-called cutoff-method. Coming from high energies, for each
delay it is determined at which energy a certain threshold n_{thr} value
is reached. Secondly, for gas reference scans, one can also calculate
the center-of-gravity of the spectrum for each delay. For this so-called
centre-of-gravity-method (COG), the energy region below 50 eV is
excluded to avoid electrons produced due to IR-photoemission.

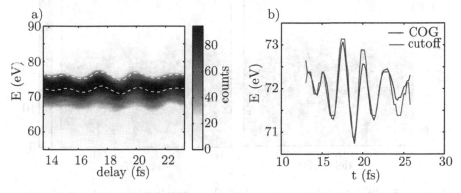

Figure 5.9: a): Smoothed gas reference streaking trace. The white
dashed lines indicate the COG-curve (lower) and cutoff-
curve (upper). b): Consistency of the two methods shifted to
a common baseline

Fig. 5.9 a) shows a gas reference scan. The white dashed lines show
the curves calculated by the cutoff-method (upper) and the COG-
method (lower). As can be seen in Fig. 5.9 b) apart from a constant
energy shift, both methods agree quite well in the determination of
the amplitudes for an appropriate value of n_{thr}, which is set to around
50% of the maximum counts. In the determination of peak positions,
the cutoff-method however is limited by the low amplitude and energy
resolution.

The cog-method can not be applied to streaking measurements from nanotips due to the electron background from scattering at lower energies nor the sophisticated analysis methods developed for streaking from plane surfaces, due to the low statistics of our measurements. In order to have a good way to analyse peak positions for tip streaking measurements, we therefore use the so-called integral method. For each delay all counts above a certain energy E_{min} are summed up. For the measurements from the nanotips E_{min} is set to lie in the cutoff region of the spectra between 86 eV and 88 eV.

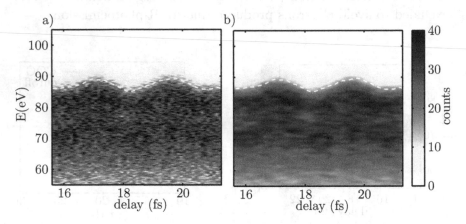

Figure 5.10: Raw (a) and smoothed (b) streaking traces obtained from a gold nanotip. The white dashed line indicates the curve obtained through the cutoff-method for the smoothed trace.

Fig. 5.10 shows a measurement for a nanotip over 4.5 hours. The relatively low statistics requires the use of spectral smoothing in order to be able to apply the cutoff-analysis method, especially for measurements with fewer counts. We employ a diskfilter which averages over directly neighbouring datapoints. In the spectrogram shown, there is no apex contribution visible. Taking the around 3000 counts in the energy region between 70 eV and 90 eV, an estimated XUV spot size of 10 μm and tip radius of 100 nm, we predict with the use of Eq. 5.8 approximately 6 electrons from the apex, for a perfect positioning of the apex within the XUV-focus. Due to the broad spectrum and the non-sharp cutoff of the electron-spectrum, the apex electrons will be completely masked by electrons from the shank. We also note that combining all measurements into one trace is not feasible due to the different experimental conditions between different measurements, like the laser waveform and intensity, which result in different streaking periods and amplitudes. Therefore we restrict our analysis to the relative amplitudes and peak shifts of gas reference and tip streaking measurements, assuming that all electrons are emitted from the shank.

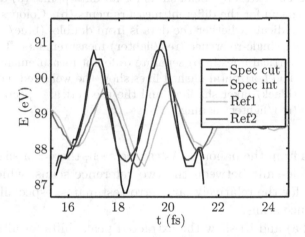

Figure 5.11: The different curves from which phase shifts and relative amplitudes are extracted. The cutoff-curve and integral-curve for the tip streaking trace as well as the cog-curves for the gas reference traces before and after the measurements. The curves have been shifted to the baseline of the cutoff-curve.

The result of such an analysis for a single measurement is shown in Fig. 5.11. The peak shifts are extracted from a comparison of the position of the peaks of integral-curves for the tip measurements and the COG-curves from the gas. Relative amplitudes are determined by examining the amplitudes from the COG-curves from gas and the

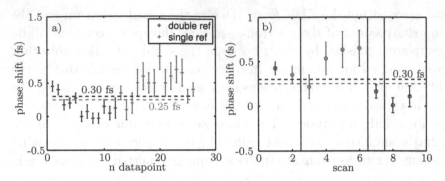

Figure 5.12: The extracted peak shifts for all data points (a) and the
mean for the different measurements (b). Colours/Shading
indicate whether the data is from double- (blue/darker)
or single-reference (red/lighter) measurements. The solid
vertical lines in b) separate different measurement days.
The horizontal dashed lines show the weighted mean
(black/daker shading) and the theoretical expectation
(red/lighter shading).

cutoff-curves from the nanotip. As can be seen, there is a slight phase
and amplitude-shift between the two reference scans, which is the
main reason for the relatively large error estimates, especially for the
single-reference scans.

Fig. 5.12 a) and b) show the extracted peak shifts for all available
data points and averaged for all measurements, respectively. The
weighted mean lies around 0.3 fs, close to the expectation of around
0.25 fs. However the data points show a huge spread, which is bigger
than the very conservative error estimates. It must therefore be concluded that there are some systematic deviations which we did not
account for so far.

By looking at Fig. 5.7 the measured values around 0.25 fs could
be explained. If the tip is not positioned appropriately, we might
measure a region of the shank with a larger radius which could lead
to the increased phase shift. However this could not explain the data
points with peak shifts above 0.5 fs and below 0.2 fs.

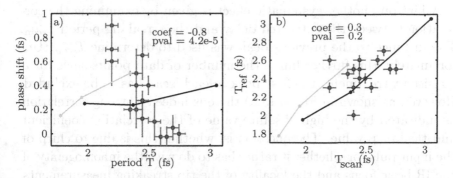

Figure 5.13: a): Correlation between the extracted peak shifts and the period of the tip streaking curve. b): Correlation between the extracted streaking periods from the gas reference and the nanotip measurements. The black/dark shaded and green/lighter shaded dots indicate the simulation results for the up-chirped and down-chirped pulses as discussed in the text. The lines a merely a guide for the eye.

Secondly there could be an influence of the Gouy phase shift in combination with a possible different positioning of the gas and tip target along the laser propagation direction. Since the count rate from the tip is so low, systematic test were performed by moving the gas target along the laser propagation direction while the TOF stayed fixed. While the count rates varied by a an order of magnitude, only a slight peak shift on the order of 100 as, which could be associated to a phase drift of the CEP stabilization, could be measured.

Furthermore the propagation of the IR-pulse through the streaking gas could lead to a phase and peak shift as has been stated in previous works [45]. This was checked by varying the pressure and consequently the count rates by one order of magnitude. Again, no significant shift was found. Thus it has to be concluded that for a fixed TOF the positioning of the gas target and the applied pressure do not influence peak positions of the gas reference scans. We note however that the gas measurement is averaging over a relatively big volume.

A hint on another systematic effect is given by examining the correlation between the extracted delay and the streaking period T_{scan}. The distance to the previous peak was used to determine T_{scan}. Unfortunately, this further limits the number of data points since for a consistent treatment, the first peak in each scan has to be excluded. However, as shown in Fig. 5.13 a) there is indeed a strong correlation as indicated by the high absolute value of the correlation coefficient and the low p-value. The question is, whether this is due to chirp of the input pulse or whether it rather has to do with the inhomogenity of the IR-laser focus and the locality of the tip streaking measurements. While for some measurements a slight chirp is directly visible for the gas reference measurement, the characterization of the tip streaking traces is difficult, since they usually only comprise two to three peaks, since they originally focused on the detection of a signal from the apex. Nevertheless a connection between chirp of the input pulse and the tip streaking can be made by looking at the correlation of the period of the gas reference and the period from the tip streaking trace. From Fig. 5.13 b) it can be seen that the correlation is relatively low from the low correlation coefficient and the high P-value. This seems to imply that the observed variation in streaking period is not due to the input pulse.

In order to understand the effect of chirp, we performed simplified calculations using a infinite cylinder model with radius of 100nm for up- and down-chirped pulse with an intensity-FWHM of 4.5 fs and chirp parameter b of $\pm\ 0.13 \cdot \frac{1}{fs^2}$. The results are indicated in Fig. 5.13 by the black/dark shaded (down-chirp) and green/lighter shaded (up-chirp) curves. As can be seen chirp leads to a change of the peak shifts, which can simply be explained by linear response theory. Due to the change of the oscillation period in time in the chirped input pulses but the constant phase shift of the linear response of the system, the simulated peak shifts should be approximately proportional to the period of the local field and the period of the input pulse and the local field should be correlated. The difference between the up- and down-chirped pulses in the plots can partly be explained by the fact that the analysis method does not determine the instantaneous frequency/period of the oscillation. It is clearly visible that the size of the spread of the measured peak shifts can not be explained by simply considering chirped input pulses. Moreover the correlation of the simulation results contradicts the experimentally observed anti-correlation.

Another possible explanation which could account for the observed spread in peak shifts and the low correlation of the periods of the gas and reference scans is the locality of the tip streaking measurements in contrast to the volume averaged gas reference measurements. Different spectral components get focussed differently which leads to a spectrally inhomogeneous field in the focal plane and, considering the double mirror, also along the laser propagation direction. While this can explain the observed spread of measured peak shifts and the low correlation of tip and gas streaking periods it is not clear whether it can explain the observed anti-correlation between peak-shifts and tip streaking periods.

There could also be some influence of surface contamination on the response of the nanotip. The presence of such contaminations can directly be seen in the electron spectrum in Fig. 5.8 a) and the missing peak of the 5p-band which is expected at around 35 eV (see e.g. [26, 40]). Indeed it is known from TEM-characterization of the nanotips that during the production a few nm thick passivation layer is formed [72]. Unfortunately simple methods exist only for the removal from the apex. Additionally our vacuum conditions are three orders of magnitude worse than in streaking experiments on flat metal surfaces, which observe surface contamination within hours [41], which implies that surface contamination seems to be unavoidable. In order to explain the observed spreads the thickness of surface contamination layers must vary considerably and always different regions of the tip shank must have been measured. We believe that our method of positioning is quite reliable, we expect that this is not the explanation for the observed effect. In the experiments different gold tips where used and not all of them were characterized directly after the measurements. The spread was however observed even in experiments performed on the same day.

We note at this point, that some confusion might arise concerning the surface contamination. From streaking measurements from tungsten surfaces covered with a monolayer of adsorbates, a shift of the relative delay between core-state-electrons and valence band electrons around 70 as could be measured [41]. Contamination layers with nm-thickness sound catastrophic in that respect. However, the nature of our measured phase shifts is completely different. They arise due to the semi-macroscopic near-field response of our nanosystem which we expect to be rather insensitive to thin surface contamination layers.

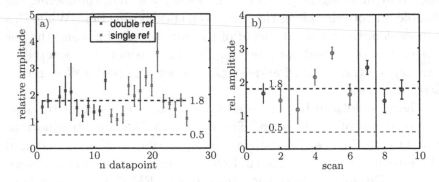

Figure 5.14: The extracted relative amplitudes for all data points (a) and the mean for the different measurements (b). Colours/Shading indicate whether the data is from double- (blue/darker) or single-reference (red/ligther) measurements. The solid vertical lines in b) separate different measurement days. The horizontal dashed lines show the weighted mean (black/darker shading) and the theoretical expectation (red/lighter shading).

The relative amplitudes between tip and gas streaking, are shown in Fig. 5.14 a) and b) for all data points and averaged for different measurements. The weighted mean lies around 1.8 and is approximately a factor of 3.5 larger as expected from simulations. For this systematic shift surface passivation layers could play a role. However, in some preliminary investigations, assuming an organic material like refractive index and varying layer thickness, this enhancement could not be explained.

Surface roughness leads to enhanced fields. This is for example used in surface-enhanced Raman-spectroscopy (SERS) [79]. From studies using Ag-surfaces with controlled roughness it is known that already a relatively low roughness on the order of few nm, could lead to the observed enhancement factors [80]. However the enhancement is not uniform in such a case and one should observe a superposition of different amplitudes. Additionally a surface roughness which produces such field enhancements should affect the phase relations. A study of this effect will be preferable, even though this might require the use of nonlocal models.

As above the most probable explanation is the of the locality of the tip streaking measurement. With our positioning method we are able to accurately place the tip in the focus. Electrons are only emitted from the surface of the tip which is confined within a small volume. The gas streaking measurements are by contrast volume averaging both along the laser propagation direction and perpendicular to it. In addition, the overlap of XUV- and IR-beam is usually not perfect, which means that for the gas streaking measurements more electrons might actually be emitted outside the IR-focus than within. We believe that the combination of those two effects is able to explain the increased streaking amplitudes from the nanotip.

The accuracy of the above analysis is limited by the relatively low amplitudes. Since the measurements were far from being affected by IR-photoemission, future measurements the amplitude could employ considerably increased amplitudes.

The measurements were performed with different tips, so the observed effects, especially the consistently too high relative amplitude, seem to be real. Nevertheless further confirmations of the above observations and tests of the possible explanations are necessary. The argument of the inhomogeneous laser-focus could be tested by scanning the tip in the plane perpendicular to the laser propagation direction with newly installed closed-loop stages. For tungsten tips a procedure exists to remove the passivation layers also from the shank [72]. Measurements before and after such a procedure could unravel the role of these layers. Furthermore, the above analysis suffers from the relatively low statistics. Improving the XUV-flux and CEP-stability will certainly be beneficial in this respect. A possibility to characterize the quality of the shank would of the tips would be helpful.

5.5 Suggestion for proof-of-principle experiment

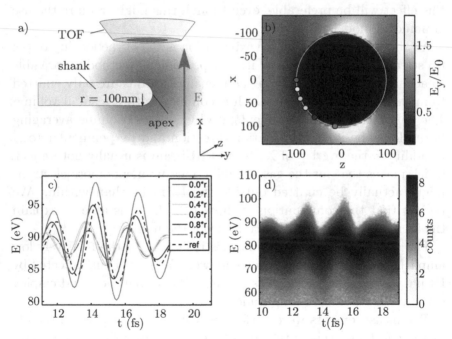

Figure 5.15: a) Geometry of the proposed experiment: The tip is rotated by 90° with respect to Fig. 5.3 b). b) The maximum field enhancement around the cylindrical nanotip shank at 800 nm away from the tip apex. The coloured points indicate the emission points for the pseudo-streaking traces (c). d) The resulting streaking trace.

As estimated from Eq. 5.8 we can not expect to see an apex contribution for our measurements. One possibility to tackle this problem is to decrease the XUV focus size. This will help in two ways. First it increases the total number of counts and furthermore improves the ratio N_{apex}/N_{shank}. However, the reduction of the focal length has proven experimentally to pose problems and the issue of having to position the apex in the focus remains. Therefore we propose a change of the geometry, with which the nanoplasmonic signature of streaking traces is in principle independent of the XUV-spotsize and with which the positioning will be easier.

The goal of a proof-of-principle experiment on nanoplasmonic streaking is to observe inhomogeneous near-fields which lead to different amplitudes and phase shifts depending on the emission position. The problem with the conventional geometry in nanoplasmonic streaking measurements is that the enhanced region is confined to a small area at the tip apex while the whole part of the shank experiences fields which lead to a uniform streaking trace. We thereforev suggest using a geometry (Fig. 5.15 a)) in which the tip is rotated by 90° compared to the conventional geometry (Fig. 5.3 a)). Due to this change the enhanced region is all along the side of the cylinder and loses its singular character, which leads to a much larger area of emission of the enhanced area. For the analysis of this geometry, we use the same methodology as in section 5.2. As can be seen in Fig. 5.15 b) this geometry leads to a region with enhanced fields at the side of the cylinder and reduced fields at the front. Additionally, there is a phase shift between the contributions from the different regions as shown in the pseudo-streaking traces (Fig. 5.15 c)). As compared to the other geometry, a significantly stronger plasmon is induced at the tip apex. However far enough away from the apex the fields still converge to the Mie-solution for infinite cylinders. The different contributions can be identified by the different amplitudes and phase shifts which are on the order of 600 as. This results in simulated streaking measurement shown in Fig. 5.15 d). All contributions a clearly visible and due to the geometry, the weight is not dependent on the XUV-spotsize. Again the influence of the change of the radius with conical tips has

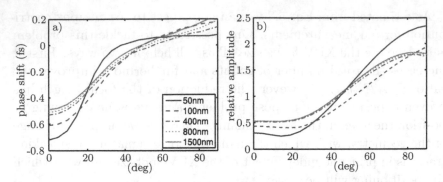

Figure 5.16: Phase shift (a) and relative amplitude (b) of the local vector potential compared to that of the input pulse. Same
as Fig. 5.7 just for laser polarization perpendicular to the
cylinder axis.

to be investigated. Using the Mie-type procedure to investigate the
near-fields in detection direction in dependence on θ yields, that even
for relatively large radii the above description remains valid. As can
be seen in Fig. 5.16 a) and b) the phase shifts and relative amplitudes
of the near-fields for radii bigger than 100 nm practically coincide. Re-
markably, the maximum field enhancement is around 1.8, and is only
slightly smaller than the field enhancement at the apex for a radius of
100 nm. This shows that in this setting not even the homogeneity of
the streaking traces is affected by the XUV focus size. Nevertheless
with a small focus one could possibly investigate the incoupling of
the plasmon at the tip apex. Furthermore, due to the high counts
from the enhanced region, we expect that one can possibly investigate
effects which go beyond our simple model such as nonlinear response
and space charge effects.

6 Conclusion and Outlook

This thesis presented the first successful experimental results of attosecond streaking measurements on nanoobjects, using a gold nanotip as target. They show that attosecond streaking can even be applied to complex systems. The observations of peak shifts and relative amplitudes slightly deviated from our model simulations. Concerning the peak shifts, the inhomogeneous laser focus was given as an explanation for the observed spread in the experiments. Non-perfect overlap between IR and XUV focus in conjunction with the argument of a local measurement with a nanotip in contrast to the volume averaged measurement of the gas target was used to explain the too high relative amplitudes. However the statistics is rather limited and future systematic measurements, which test those hypotheses, are desirable.

This thesis shows that within a semiclassical description the neglect of electron transport has only a minor effect on the time the electrons need to reach the surface, which is due to the small inelastic mean free paths. A consideration effect will only become important when measurement with very low statistical and systematic uncertainties are available. Concerning the timing, the neglect of electron transport used in previous studies therefore seems to be justified.

Our model however predicts a strong decrease in photoelectron yield with respect to the surface emission angle. This effect might become important when the nanobject posseses cross sectional areas of equal size with different near-fields, such as spheres or the cylindrical geometry suggested in the last section.

An original goal of this thesis was to measure electrons from the enhanced near-field region at the tip apex. This was not achieved, which could be explained with the help of a simple count rate approximation. Simulations suggest that a reduction of XUV spotsize will enable such measurements. Alternatively in this thesis, a slight change of geometry is suggested, which will possibly yield a clear signature of nanoplasmonic streaking without the need of extremely small XUV spot sizes. Attosecond streaking measurement with this geometry will also be a test for our model and possibly even allow the study of higher order effects, such as nonlinearities and space charge effects.

The experimental results presented in this thesis lead to a paper, which is currently waiting for publication [31].

Appendix A: Description of electron scattering

The electron scattering inside the bulk is determined by the total mean free path :

$$\frac{1}{\lambda_{tot}} = \frac{1}{\lambda_{el}} + \frac{1}{\lambda_{el}} \tag{1}$$

The probability p(s) that an electron scatters in the solid at a distance s from it's current position is given by:

$$p(s) = \frac{1}{\lambda_{tot}} e^{-\frac{s}{\lambda}} \tag{2}$$

Consequently using the inversion cumulative distribution function, s following the statistics given by the above equation can be computed from a uniform probability distribution P in the interval [0,1]:

$$s = -\lambda_{tot} \log(1 - P) \tag{3}$$

Using the velocity v of the electron this can also be converted to a scattering time $\tau = s/v$. Note that this formulation makes the assumption that the electron velocity is constant. For the electrons inside solids in an attosecond streaking experiment this is only approximately true. If the electron has crossed the surface before travelling the distance s, no bulk scattering occurs. Otherwise the probability of an elastic or inelastic scattering event, P_{el} and P_{inel} can be computed from the ratios of the inverse mean free paths:

$$P_{el} = \frac{\lambda_{inel}}{\lambda_{el} + \lambda_{inel}} \tag{4}$$

Using again a uniform random number distribution P, an elastic scattering event occurs if $P < P_{el}$ else an inelastic scattering event occurs.

All the other distribution functions of elastic and inelastic scattering follow more or less complex curves and the inverse cumulative distribution function is not known analytically. Sampling the distributions function at closely spaced positions the cumulative distribution function can numerically be calculated. Inversion is then achieved by switching position vector and cumulative distribution function. Again using linear interpolation the inverse cumulative distribution function can now be calculated. For multidimensional distribution functions, the dimension is subsequently reduced by calculating conditional inverse cumulative distribution functions. For a given random variable the conditional inverse cumulative distribution function is then calculated by multidimensional linear interpolation. In this approach only a single random number has to be generated per dimension involved in the distribution function however at the price of increased memory requirements. However due to the smooth change of the distribution functions involved in the description of scattering in solids a database of a few 100 Mb yielded a good sampling of the distribution functions. For an inelastic scattering event using a uniform random number distribution first the energy loss Q is calculated and then for the given Q the scattering angle θ. Only the scattering angle θ has to be calculated in an elastic scattering event.

When the electron reaches the surface it has to cross the surface barrier and may undergo surface scattering. According to the description in chapter 3 the surface scattering is symmetrically separated into a part before and after the surface transmission. Due to the Poissonian statistics of the surface scattering the total SEP therefore has to be divided by a factor of 2. If the electron undergoes scattering, one scattering event is calculated. Then for the new angle and energy this procedure is maximally repeated 5 times until no scattering occurs. Then the electron is diffracted at at surface barrier and weighted by the transmission factor. Finally the other part of the surface scattering is calculated analogously to the first part.

The splitting of the surface scattering into two equal parts is questionable, but it was found to be more consistent when including the

transmission. Also the possibility to switch off the surface scattering
is implemented.

Bibliography

[1] Herb Sutter. The free lunch is over: A fundamental turn toward concurrency in software. *Dr. Dobb's Journal*, 30, 205. graph updated in 2009 taken from: "http://www.gotw.ca/publications/concurrency-ddj.htm"; last accessed 08.06.2014 Online.

[2] Ferenc Krausz and Mark I. Stockman. Attosecond metrology: from electron capture to future signal processing. *Nature Photonics*, 8:205 213, 2014.

[3] E. Goulielmakis, V. S. Yakovlev, A. L. Cavalieri, M. Uiberacker, V. Pervak, A. Apolonski, R. Kienberger, U. Kleineberg, and F. Krausz. Attosecond control and measurement: Lightwave electronics. *Science*, 317:769–775, 2007.

[4] Martin Schultze, Elisabeth M. Bothschafter, Annkatrin Sommer, Simon Holzner, Wolfgang Schweinberger, Markus Fiess, Michael Hofstetter, Reinhard Kienberger, Vadym Apalkov, Vladislav S. Yakovlev, Mark I. Stockman, and Ferenc Krausz. Controlling dielectrics with the electric field of light. *Nature*, 493:75–78, 2013.

[5] Agustin Schiffrin, Tim Paasch-Colberg, Nicholas Karpowicz, Vadym Apalkov, Daniel Gerster, Sascha Mühlbrandt, Michael Korbman, Joachim Reichert, Martin Schultze, Simon Holzner, Johannes V. Barth, Reinhard Kienberger, Ralph Ernstorfer, Vladislav S. Yakovlev, Mark I. Stockman, and Ferenc Krausz. Optical-field-induced current in dielectrics. *Nature*, 493:70–74, 2013.

[6] Mark I. Stockman. Nanoplasmonics: past, present, and glimpse into future. *Optics Express*, 19:22029, 2011.

[7] Stefan A. Maier. *Plasmonics: Fundamentals and Applications*. Springer, New York, 2007.

[8] Mark I. Stockman. Nanoplasmonics: The physics behind the applications. *Phys. Today*, 64:3944, 2011.

[9] Markus Schenk, Michael Krüger, and Peter Hommelhoff. Strong-field above-threshold photoemission from sharp metal tips. *Review of Scientific Instruments*, 82:026101, 2011.

[10] C. Ropers, C. C. Neacsu, T. Elsaesser, M. Albrecht, M. B. Raschke, and C. Lienau. Grating coupling of surface plasmons onto metallic tips: a nano-confined light source. *Nano Lett.*, 7:2784–2789, 2007.

[11] S. Berweger, J. M. Atkin, X. G. Xu, R. L. Olmon, and M. B. Raschke. Femtosecond nanofocusing with full optical waveform control. *Nano Lett.*, 11:4309–4313, 2011.

[12] F. De Angelis, G. Das, P. Candeloro, M. Patrini M. Galli, A. Bek, M. Lazzarino, I. Maksymov, C. Liberale, and L. C. Andreani and. Di Fabrizio. Nanoscale chemical mapping using three-dimensional adiabatic compression of surface plasmon polaritons. *Nature Nanotechnology*, 5:67–72, 2010.

[13] Diyar Sadiq, Javid Shirdel, Jae Sung Lee, Elena Selishcheva, Namkyoo Park, , and Christoph Lienau. Adiabatic nanofocusing scattering-type optical nanoscopy of individual gold nanoparticles. *Nano Lett.*, 11:16091613, 2011.

[14] Catalin C. Neacsu, Samuel Berweger, Robert L. Olmon, Laxmikant V. Saraf, Claus Ropers, and Markus B. Raschke. Near-field localization in plasmonic superfocusing: A nanoemitter on a tip. *Nano Lett.*, 10:592–596, 2010.

[15] M. Krüger, M. Schenk, and P. Hommelhoff. Attosecond control of electrons emitted from a nanoscale metal tip. *Nature*, 475:78–81, 2011.

[16] G. Wachter, Ch. Lemell, J. Burgdörfer, M. Schenk, M. Krüger, and P. Hommelhoff. Electron rescattering at metal nanotips induced by ultrashort laser pulses. *Phys. Rev. B*, 86:035402, 2012.

[17] Björn Piglosiewicz, Slawa Schmidt, Doo Jae Park, Jan Vogelsang, PetraGroß, Cristian Manzoni, Paolo Farinello, Giulio Cerullo, and Christoph Lienau. Carrier-envelope phase effects on the strong-field photoemission of electrons from metallic nanostructures. *Nature Photonics*, 8:37–42, 2014.

[18] J. Itatani, F. Quere, G. L. Yudin, M. Yu. Ivanov, F. Krausz, and P. B. Corkum. Attosecond streak camera. *Physical Review Letters*, 88:173903, 2002.

[19] Markus Kitzler, Nenad Milosevic, Armin Scrinzi, Ferenc Krausz, and Thomas Brabec. Quantum theory of attosecond xuv pulse measurement by laser dressed photoionization. *Physical Review Letters*, 88:173904, 2002.

[20] R. Kienberger, E. Goulielmakis1, M. Uiberacker, A. Baltuska, V. Yakovlev1, F. Bammer, A. Scrinzi, Th. Westerwalbesloh, U. Kleineberg, U. Heinzmann, M. Drescher, and F. Krausz. Atomic transient recorder. *Nature*, 427:817–821, 2004.

[21] M. Schultze, M. Fieß, N. Karpowicz, J. Gagnon, M. Korbman, M. Hofstetter, S. Neppl, A. L. Cavalieri, Y. Komninos, Th. Mercouris, C. A. Nicolaides, R. Pazourek, S. Nagele, J. Feist, J. Burgdrfer, A. M. Azzeer, R. Ernstorfer, R. Kienberger, U. Kleineberg, E. Goulielmakis, F. Krausz, and V. S. Yakovlev. Delay in photoemission. *Science*, 328:1658–1662, 2010.

[22] A. L. Cavalieri, N. M üller, T. Uphues, V. S. Yakovlev, A. Baltuska, B. Horvath, B. Schmidt, L. Blümel, R. Holzwarth, S. Hen-

del, M. Drescher, U. Kleineberg, P. M. Echenique, R. Kienberger, F. Krausz, and U. Heinzmann. Attosecond spectroscopy in condensed matter. *Nature*, 449:1029–1032, 2007.

[23] A. Cavalieri, F. Krausz, R. Ernstorfer, R. Kienberger, P. Feulner, J. Barth, and D. Menzel. *Attosecond Time-Resolved Spectroscopy at Surfaces. In Dynamics at Solid State Surfaces and Interfaces.* 2010.

[24] S. Neppl, R. Ernstorfer, E. M. Bothschafter, A. L. Cavalieri, D. Menzel, J. Barth, J. V., F.Krausz, R. Kienberger, and P. Feulner. Attosecond time-resolved photoemission from core and valence states of magnesium. *Physical Review Letters*, 109:087401, 2012.

[25] Frederik Süßmann and Matthis F. Kling. Attosecond nanoplasmonic streaking of localized fields near metal nanospheres. *Phys. Rev. B*, 84:241–248, 2011.

[26] E. Skopalova, D. Y. Lei, T. Witting, C. Arrell, F. Frank, Y. Sonnefraud, S. A. Maier, J. W. G. Tisch, and J. P. Marangos. Numerical simulation of attosecond nanoplasmonic streaking. *New J. Phys.*, 13:083003, 2011.

[27] F. Kelkenberg, A. F. Koenderink, and M. J. J. Vrakking. Attosecond streaking in a nano-plasmonic field. *New J. Phys.*, 14:093034, 2012.

[28] A. G. Borisov, P. M. Echenique, and A. K. Kazansky. Attostreaking with metallic nano-objects. *New J. Phys.*, 14:23036, 2012.

[29] James S. Prell, Lauren J. Borja, Daniel M. Neumark, and Stephen R. Leone. Simulation of attosecond-resolved imaging of the plasmon electric field in metallic nanoparticles. *Ann. Phys.*, 535:151–161, 2013.

[30] Mark I. Stockman, Matthias F. Kling, Ulf Kleineberg, and Ferenc Krausz. Attosecond nanoplasmonic-field microscope. *Nature Photonics*, 1:539, 2007.

[31] Frederik Suessmann Michael Foerster Michael Krueger Byung-Nam Ahn Karen Wintersperger Sergey Zherebtsov Alexander Guggenmos Vladimir Pervak Alexander Kessel Sergei Trushin Abdallah Azzeer Mark Stockman Dong-Eon Kim Ferenc Krausz Peter Hommelhoff Benjamin Frg, Johannes Schoetz and Matthias Kling. Attosecond nanoscale near-field sampling. *arXiv:1508.05611v1 [physics.optics]*, 2015.

[32] Jean-Claude Diels and Wolfgang Rudolph. *Ultrashort Laser Pulse Phenomena*. Elsevier Inc., Burlington, 2006.

[33] Uwe Thumm, Qing Lao, Elisabeth Bothschafter, Frederik Süßmann, Matthias F. Kling, and Reinhard Kienberger. *Handbook of Photonics*, chapter Attosecond Physics: Attosecond Streaking Spectroscopy of Atoms and Solids. Wiley-Blackwell. to be published.

[34] Pierre Berini, Alexandre Bouhelier, Javier Garcia de Abajo, and Namkyoo Park. Focus issue on surface plasmon photonics introduction. *Optics Express*, 21:27286 27290, 2013.

[35] Oliver Benson. Assembly of hybrid photonic architectures from nanophotonic constituents. *Nature*, 480:193199, 2011.

[36] Craig F. Bohren and Donald R. Huffman. *Absorption and Scattering of Light by Small Particles*. John Wiley and Sons, New York, 1983.

[37] Michael Quinten. *Optical Properties of Nanoparticle Systems: Mie and Beyond*. Wiley-VCH Verlag, Weinheim, 2011.

[38] Zenghu Chang. *Fundamentals of Attosecond Optics*. CRC Press, Boca Raton, 2011.

[39] *Solid-State Photoemission and Related Methods: Theory and Experiment*, chapter Overview of core and valence photoemission. Wiley VCH, 2003.

[40] Friedrich Reinert and Stefan Hüfner. Photoemission spectroscopy - from early days to recent applications. *New J. Phys.*, 7:97, 2005.

[41] Stefan Neppl. *Attosecond Time-Resolved Photoemission from Surfaces and Interfaces*. PhD thesis, TU München, 2011.

[42] C. Lemell, B. Solleder, K. Tökesi, and J. Burgdörfer. Simulation of attosecond streaking of electrons emitted from a tungsten surface. *Physical Review A*, 79:062901, 2009.

[43] The electronic structure project, 2012. [Online; accessed 2-June-2014].

[44] C. Ortiz, O. Eriksson, and M. Klintenberg. Data mining and accelerated electronic structure theory as a tool in the search for new functional materials. *Comput. Mater. Sci.*, 44:1042–1049, 2009.

[45] Frederik Süßmann. *Attosecond dynamics of nano-localized light fields*. PhD thesis, LMU München, 2013.

[46] Andrius Baltuska, Matthias Uiberacker, Eleftherios Goulielmakis, Reinhard Kienberger, Vladislav S. Yakovlev, Thomas Udem, Theodor W. Hnsch, , and Ferenc Krausz. Phase-controlled amplification of few-cycle laser pulses. *IEEE JOURNAL OF SELECTED TOPICS IN QUANTUM ELECTRONICS*, 9:972–989, 2003.

[47] Albert Messiah. *Quantum Mechanics*. Interscience Publishers, New York, 1962.

[48] Aleksander Jablonski, Francesc Salvat, and Cedric J. Powell. Comparison of electron elastic-scattering cross sections calculated from two commonly used atomic potentials. *J. Phys. Chem. Ref. Data*, 33:409–451, 2004.

[49] Francesc Salvat, Aleksander Jablonski, and Cedric J. Powell. Elsepa dirac partial-wave calculation of elastic scattering of electrons and positrons by atoms, positive ions and molecules. *Computer Physics Communications*, 165:157–190, 2005.

[50] G. Wachter, C. Lemell, J. Burgdörfer, M. Schenk, M. Krüger, and P.Hommelhoff. Electron rescattering at metal nanotips induced by ultrashort laser pulses. *Phys. Rev. B*, 86:035402–035406, 2012.

[51] Francesc Salvat. Optical-model potential for electron and positron elastic scattering by atoms. *Physical Review A*, 68, 2003.

[52] H.-J. Fitting, E. Schreiber, J.-Chr., and A. von Czarnowski. Attenuation and escape depths of low-energy electron emission. *Journal of Electron Spectroscopy and Related Phenomena*, 119:3547, 2001.

[53] *Surface and Interface Science*, chapter An Introduction to the Theory of Crystalline Elemental Solids and their Surfaces. Wiley VCH, 2012.

[54] Wolfgang S. M. Werner and Francesc Salvat-Pujol. Surface excitations in electron spectroscopy. part i: dielectric formalism and monte carlo algorithm. *Surface Interface Analysis*, 45:873–894, 2012.

[55] D. Emfietzoglou, I. Kyriakou1, I. Abril, R. Garcia-Molina, and H. Nikjoo. Inelastic scattering of low-energy electrons in liquid water computed from optical-data models of the bethe surface. *International Journal of Radiation Biology*, 88:22–28, 2012.

[56] Wolfgang S. M. Werner, Kathrin Glantschnig, and Claudia Ambrosch-Draxl. Optical constants and inelastic electron scattering data for 17 elemental metals. *Journal of Physics: Chem. Ref. Data*, 38:1013–1092, 2009.

[57] N. D. Mermin. Lindhard dielectric function in the relaxation -time approximation. *Physical Rev. B*, 1:2362–2363, 1970.

[58] V. Cataudella, V. Marigliano Ramagua, and G.P. Zucchelli. On the analytical struture of the lindhard dielectric function. *Physics Letters*, 92A:359–362, 1982.

[59] N. D. Mermin. Lindhard dielectric function in the relaxation-time approximation. *Physical Review B*, 1:2362–2363, 1970.

[60] Z.-J. Ding and R. Shimizu. Inelastic collisions of kev electrons in solids. *Surface Science*, 222:313–331, 1989.

[61] D. Emfietzoglou, I. Abril, R. Garcia-Molina, I.D. Petsalakis, H. Nikjoo, I. Kyriakou, and A. Pathak. Semi-empirical dielectric descriptions of the bethe surface of the valence bands of condensed water. *Nuclear Instruments and Methods in Physics Research B*, 266:11541161, 2008.

[62] Edward D. Palik. *Handbook of Optical Constants of Solids*. Academic Press, Boston, 1985.

[63] C. J. Tung, Y. F. Chen, C. M. Kwei, and T. L. Chou. Differential cross sections for plasmon excitations and reflected electron-energy-loss spectra. *Physical Review B*, 49:16684–16693, 1994.

[64] J.-Ch. Kuhr and H.-J. Fitting. Monte-carlo simulation of low energy electron scattering in solids. *phys. stat. sol.*, 172:433–449, 1999.

[65] D. Emfietzoglou, I. Kyriakou, I. Abril, R. Garcia-Molina, I.D. Petsalakis, H. Nikjoo, and A. Pathak. Electron inelastic mean free paths in biological matter based on dielectric theory and local-field corrections. *Nuclear Instruments and Methods in Physics Research B*, 267:4552, 2009.

[66] J.-Ch. Kuhr and H.-J. Fitting. Monte carlo simulation of electron emission from solids. *J. Phys. Chem. Ref. Data*, 28:19–61, 1999.

[67] C. J. Powell and A. Jablonski. Nist electron inelastic-mean-free-path database. Technical report, National Institute of Standards and Technology NIST, 2010.

[68] Dimitris Emfietzoglou, Ioanna Kyriakou, Rafael Garcia-Molina, and Isabel Abril. The effect of static many-body local-field corrections to inelastic electron scattering in condensed media. *Journal of Applied Physics*, 114:144907, 2013.

[69] Michael Bosman et al. Surface plasmon damping quantified with an electron nanoprobe. *Scientific Reports*, 3:1312, 2013.

[70] F. J. Garcia de Abajo. Optical excitations in electron microscopy. *Rev. Mod. Phys.*, 82:209, 2010.

[71] Wolfgang S. M. Werner, Werner Smekal, and Francesc Salvat-Pujol. Angular dependence of electron induced surface plasmon excitation. *Applied Physics Letters*, 98, 2011.

[72] Michael Krüger. *Attosecond Physics in Strong-Field Photoemission from Metal Nanotips*. PhD thesis, LMU München, 2013.

[73] Lumerical solutions, inc.

[74] P. B. Johnson and R. W. Christy. Optical constants of the noble metals. *Phys. Rev. B*, 6:4370–4379, 1972.

[75] Sebastian Thomas, Michael Krüger, Michael Förster, Markus Schenk, and Peter Hommelhoff. Probing of optical near-fields by electron rescattering on the 1 nm scale. *Nano Lett.*, 13:47904794, 2013.

[76] Loup Verlet. "computer "experiments" on classical fluids. i. thermodynamical properties of lennardjones molecules". *Physical Review*, 159:98–103, 1967.

[77] J. Schäfer, S.-C. Lee, and A. Kienle. Calculation of the near fields for the scattering of electromagnetic waves by multiple infinite cylinders at perpendicular incidence. *J. Quant. Spectrosc. Radiat. Trans.*, 113, 2012.

[78] C. H. Zhang and U. Thumm. Probing dielectric-response effects with attosecond time-resolved streaked photoelectron spectroscopy of metal surfaces. *Physical Review A*, 84:063403, 2011.

[79] M. Moskovits. surface-enhanced spectroscopy. *Reviews of Modern Physics*, 57:783–826, 1985.

[80] Yuda Zhao, Xin Liu, Dang Yuan Lei, and Yang Chai. Effects of surface roughness of ag thin films on surface-enhanced raman spectroscopy of graphene: spatial nonlocality and physisorption strain. *Nanoscale*, 6:1311–1317, 2014.

Printed in the United States
By Bookmasters